THE EUROPEAN MARKET FOR FRUIT AND VEGETABLES

THE EUROPEAN MARKET FOR FRUIT AND VEGETABLES

LYNN HINTON

*Assistant Director of Research,
Department of Land Economy,
Cambridge University, UK*

*Fellow of Wolfson College,
Cambridge, UK*

ELSEVIER APPLIED SCIENCE
LONDON and NEW YORK

ELSEVIER SCIENCE PUBLISHERS LTD
Crown House, Linton Road, Barking, Essex IG11 8JU, England

Sole Distributor in the USA and Canada
ELSEVIER SCIENCE PUBLISHING CO., INC.
655 Avenue of the Americas, New York, NY 10010, USA

WITH 94 TABLES AND 40 ILLUSTRATIONS

British Library Cataloguing in Publication Data
Hinton, Lynn
The European market for fruit and vegetables
I. Title
338.174094

ISBN 1-85166-662-1

Library of Congress CIP data applied for

Printed in Great Britain at The Alden Press Ltd, Oxford

FOREWORD

Lynn Hinton has undertaken the daunting task of making sense out of the dynamic and diverse market for fresh fruits and vegetables in the European Community. He achieved this goal because he concentrated on determining the facts that accurately describe the market, an area where most marketing disputes begin and end. The information provided here allows attention to shift to where it ought to be: on the implications these facts have for private and public strategies. The book should be a useful reference for all those concerned with this large market.

Although the production of fruit and vegetables in the EEC is dwarfed by the production of cereals, livestock and other basic crops, it still remains a big business. Production is over 74 million metric tonnes annually and total human consumption amounts to 68 million metric tonnes. The Community is on a par with the United States as an exporter of fruits and vegetables, but is a significantly larger importer. The dynamics of the market is expressed by the emergence of new varieties of fresh produce, changed consumer preferences, altered channels of distribution, and new rules of the game. The diversity of the market rests on the multitude of fruits and vegetables available throughout the year, the numerous production sources for them, and the complex policies that influence their consumption.

Trade policy affecting fruit and vegetables received considerable attention in the Uruguay Round of the GATT negotiations and in the debate on the European Community's move to a single market in 1992. It has also been the focus of bilateral disputes between the Community, the United States and Japan, among others. The author has walked a tightrope between the opposing views on trade policy, and while I might think that he has been too harsh on the United States and too gentle on the Community, the end result is a reasonably balanced view of a very complex situation.

The real issues concern non-tariff barriers to trade rather than tariff barriers. Unfortunately, the world has raised non-tariff barriers to counteract tariff concessions. Japan, the United States and the Community all use systems of quotas, sanitary requirements, and product standards to protect domestic producers. As Lynn Hinton points out, it is the degree of protection that is at question. Most economists believe that production, trade and consumption would be best served if markets for fruit and vegetables were free from any barriers other than those needed to protect health and safety. However, this idealised world is not the world in which we live. Political and social compromises tend to alter free markets and to foster strong constituencies for continued government intervention. Until this is unravelled, policy will be a major factor in the European market for fruit and vegetables.

There are several important export and import countries within the Community that strongly influence market results. The author analyses the factors within these countries that explain the production or trade that we all observe. He first examines West Germany (now the western states of the newly united Germany). West Germany imports more fruit and vegetables than any other country in the world. Its relatively large and wealthy population has a preference for fresh produce, and that preference cannot be satisfied by German farmers, hampered as they are by cool and humid weather. German consumers demand fresh produce throughout the year, thus stimulating imports, and they seek new and different varieties of fresh products, sometimes at the expense of processed products. Germany has developed an efficient food distribution system which may well reach into other countries as European integration proceeds.

The United Kingdom is Lynn Hinton's home territory and his understanding of the market is evident. He introduced the market to me in 1984 and it increased my perspective of global competition significantly. One can really identify the successful operators in the UK market, such as the Netherlands, Spain or Israel, by visiting their distribution facilities or sales offices. The UK market is truly diverse and maintains a high demand for traditional and exotic fruits and vegetables. It relies heavily on domestic producers for its vegetable requirements, but looks overseas for much of its fruit. Lynn Hinton is optimistic about future expansion of the market as British consumers shift toward more healthy diets containing larger amounts of fresh fruits and vegetables.

One of the most interesting chapters concerns the Netherlands which is a major exporter of vegetables and fruit. Surprising as this may be for a small country in the northern temperate zone, it is surpassed by the amazingly efficient Dutch auction system that makes the country a major marketing and redistribution centre for both domestic and imported fresh produce. Lynn Hinton has done an excellent job in explaining why the Netherlands is so important to the Community's production of fresh produce and in detailing the development and operation of auction markets. It is clear that the author has studied the subject carefully.

The Mediterranean exporters are Greece, Italy, Spain and Portugal. If one can say that there is a "bread basket" for the Community's supply of vegetables and fruit, these countries would constitute that basket. Yet, as the author correctly points out, each of these countries has a different production and distribution system that leads to different export practices. They should not be considered as homogeneous for analytical or policy purposes. The implementation of the Common Agricultural Policy has helped bridge the gap between these producers but it is unlikely to bring uniformity. The development of aggressive modern marketing organisation is doing much to capitalise on the unique advantages that exist in each country.

In the concluding chapter, the author provides a needed review of global marketing opportunities for developing countries. He stresses the need for improving infrastructure and production and management skills. This corresponds exactly with my observations after a decade of interviewing managers throughout the world. The needed changes will require improved research and extension systems and far better market information networks. Significant export opportunities will be realised only if developed countries, specifically the European Community, Japan and the United States reduce their market barriers. Steps have been made through a wide variety of special trade measures but more remains to be done - a sad legacy of the GATT trade negotiations. Within the Community, the drive to a more integrated internal market, allows lower prices and stimulates increased demand. The result may well be increased potential for all suppliers to the European market for fruit and vegetables.

Kirby Moulton
University of California,
Berkeley
April 1, 1991

Author's Note and Acknowledgements

Lynn Hinton entered the field of horticultural economics at Wye College (University of London), where he was appointed Lecturer in Agricultural Economics with special reference to the horticultural sector. Wye College was especially strong in this field of research at that time, with three researchers, while other university centres of agricultural economics had only one or none.

He then moved to Cambridge University and developed research in the horticultural sector there, a subject not followed at Cambridge before. At first the emphasis was on horticultural management and marketing policy. In the early 1960s a number of Bank Review papers were published, including Capital Needs in Horticulture, Policy for Horticulture and The Economic Contribution of Horticulture. His first paper on Horticulture and the Common Market was published in the Journal of the Agricultural Economics Society in 1968, and a number of papers on the fruit and vegetable sector of the EEC have been published since then. These include the European Fruit and Vegetable Sector (1976), while on sabbatical leave at the University of Munich, EEC Enlargement in the Fruit and Vegetable Sector with Special Reference to Spain and the United Kingdom (1983), Spanish Accession to the EEC: Prospects for Temperate fruit and Vegetables (1984), EEC Enlargement in the Fruit and Vegetable Sector (1984) and Enlargement of the European Economic Community in the Fruit and Vegetable Sector (1986). Over this time he was appointed as a Research Officer at Cambridge, becoming Assistant Director of Research in 1968. He has visited all member states of the EEC except Greece and Portugal.

Many of the references for this book relate to the proceedings of the Economics Commission of the International Society for Horticultural Science, to which the writer has been a prominent subscriber. He is an Editorial collaborator for International Fruit World, an Executive Committee member of the UK Agricultural Economics Society, and a Fellow of the Institute of Horticulture.

The author is grateful for the assistance of Mrs. W. Champernowne and Mrs. Anne Davenport in translating Dutch and German papers, to Mrs. Carol Asby, who assisted with the statistical analysis, and to Mrs. Joy Meyrick, who typed the manuscript. Assistance was also received from Mme. A-M. de Pascale, Head of the Horticulture Division of the Commission of the European Communities, and Mr. R. Muthmann of the Statistical Office of the European Communities.

Data on imports and exports are taken from Eurostat External Trade - Imports and Exports for 1989, which gives figures for 1987. Figures for areas and production of fruit and vegetables also come from Eurostat 1989, giving figures for 1987. Data for West Germany is substantially from publications of the Zentral Markt und Preisberichstelle 1987-88, and for Holland from publications of the Dutch Bureau of Auctions. For Holland much of the data relates to 1988.

Cambridge, March 1991 Lynn Hinton

TABLE OF CONTENTS

CHAPTER 1

AN INTRODUCTION TO EUROPEAN TRADE IN FRUIT AND VEGETABLES

The Impact of the Trade in Fruit and Vegetables

Various factors contribute to make the European Community the biggest world market for fruit and vegetables, despite her large domestic production. Indeed, until the EEC enlargement of 1986 to include Spain and Portugal, the European Economic Community was only some 40 per cent self-sufficient for citrus, but enlargement has brought more self-sufficiency in this sector. International trade in fruit and vegetables is a function of climate and of season. Different climates from those in Europe occasion the importation of exotic fruit and vegetables, and different seasons bring fresh produce into Europe in the off- season of home supply, while the climatic range within Europe accounts for an important intra-Community trade. Europe is the largest international market because, unlike the United States, she has fewer production regions of such contrasting climate. High consumption in Europe stems from a good supply of traditional domestic produce in all countries of the EEC, and it also stems from Europe's traditional links with overseas contacts, and of course from high incomes in Europe.

Of particular relevance is the fruit and vegetable policy of the EEC which has protective régimes for a number of fruit and vegetable commodities. This high protection has given rise to serious disputes, particularly in West Germany, between the trade and the associations of producers, and no doubt some measure of consumer choice is interfered with by the EEC policy for fruit and vegetables, while it also may be contended that the vexed question of fruit and vegetable surpluses also arises in part because of EEC policy. It is important for the developing world that their competitive fruit and vegetable commodities should not be penalised because of EEC protection against them.

There is no doubt about the health and enjoyment features of fruit and vegetables. As regards health the United Kingdom food Survey for 1987 stated that, nutritionally, fruit and vegetables are important sources of a number of nutrients most notably vitamin C and to a lesser extent vitamin A (as Carotene), thiamine and iron. The relative importance of particular fruit and vegetables as sources of vitamin C has reversed since 1965 with the decline in potato consumption and the marked increase in fruit juice consumption. The reduced potato consumption has also contributed to lower total intakes of iron. There has been little change in the amount of vitamin A and thiamine provided by fruit and vegetables in the household diet since 1965. Fruit juice now provides more than three times as much vitamin C as potatoes which until 1985 were the main source of vitamin C in the household diet.[1]

---

[1] Household Food Consumption and Expenditure 1987 HMSO 1989.

Table 1. World Exports and EEC External Trade 1973-1987 - million US dollars

	1973	1980	1981	1982	1983	1984	1985	1986	1987p
WORLD EXPORTS									
All products	451	1604	1628	1513	1482	1575	1573	1673	1928
of which									
Agricultural products	**101**	**244**	**243**	**221**	**218**	**230**	**216**	**230**	**257**
Other products	351	1361	1385	1292	1264	1345	1357	1436	1671
EXTERNAL EEC TRADE									
Exports									
All products	96	305	297	279	270	279	289	336	392
of which									
Agricultural products	**10**	**29**	**31**	**26**	**25**	**26**	**26**	**28**	**33**
Imports									
All products	108	394	353	328	305	311	310	329	393
of which									
Agricultural products	**31**	**63**	**54**	**50**	**48**	**49**	**47**	**52**	**59**
WORLD EXPORTS of Agricultural products as percentage of world exports	**22**	**15**	**15**	**15**	**15**	**15**	**14**	**14**	**13**
EEC EXPORTS of Agricultural products as percentage of total EEC exports	**10**	**9**	**10**	**10**	**9**	**9**	**9**	**8**	**8**
EEC IMPORTS of Agricultural products as percentage of total EEC imports	**29**	**16**	**15**	**15**	**16**	**16**	**15**	**16**	**15**
INDEX CHANGES									
EEC exports of agricultural products									
- base 1973 = 100	**100**	**318**	**309**	**290**	**281**	**290**	**301**	**350**	**406**
- base 1980 = 100		**100**	**107**	**92**	**87**	**91**	**91**	**98**	**114**
EEC imports of agricultural products									
- base 1973 = 100	**100**	**300**	**321**	**275**	**259**	**273**	**272**	**295**	**334**
- base 1980 = 100		**100**	**85**	**80**	**76**	**77**	**74**	**83**	**94**

p = provisional

Source: GATT Statistics, Eurostat 1988 and The Agricultural Situation in the Community 1988 Report. Brussels 1989.

European Trade in Fruit and Vegetables in the Context of Agricultural and World Trade

The European Community, on account of the range of climate within it, has a pronounced internal trade in fruit and vegetables as well as a very significant international trade. In a Community of 10, production of fruit and vegetables represented 12 per cent of EEC agricultural output, but with the accession of Spain and Portugal in 1986 this proportion increased, as in Spain alone, these products constitute some 20 per cent of that country's final agricultural output.[2]

Before dealing with EEC trade in fruit and vegetables, this sector of trade is put into perspective with world and EEC trade as a whole and agricultural trade. As shown in Table 1 world exports of all products in 1987 reached a total of some 1.9 billion dollars, of which EEC agricultural exports are 17 per cent. This is in contrast to 1973 when the proportion of agricultural exports of total world exports was 22 per cent. The decline of agricultural exports is a feature of the present trend in world exports as a whole. However, the EEC share of total world exports maintained a near constant of a little over 20 per cent over the period 1973 to 1987, and EEC agricultural exports fell slightly from some ten to somewhat over eight per cent of EEC exports of all products.

Since 1973 there has been a continuous decline of world exports of agricultural products as a proportion of total world exports, from 22.4 per cent in 1973 to 13.3 per cent in 1987. EEC agricultural exports as a proportion of total EEC exports has also declined over the period from ten per cent to 8.4 per cent. EEC imports of agricultural products as a proportion of total EEC imports, 28.7 per cent in 1973, have remained at a fairly constant level of around 15 per cent from 1980 to 1987.

The indices of EEC exports of agricultural products and EEC imports of agricultural products are given on two bases, 1973 and 1980. While there was a big change from 1973 to 1980, the export index from 1980 to 1987 shows a fluctuating but fairly regular trend. The export index has a high point in 1987 of 113.9 and a low point of 86.5 in 1983. The index of agricultural imports runs between 74 and 94. It shows a high point in 1987 and a low point in 1985.

Table 2. EEC Agricultural Exports and Imports, 1980 and 1986, Compared with the USA - million dollars

	EEC 12		USA	
	1980	1986	1980	1986
ALL EXPORTS	305041	336350	216592	206408
of which				
Agricultural and food products	28750	28334	46402	31207
Food products and live animals	19869	18516	27655	17152
Fruit and vegetables	2444	2622	2922	2649
ALL IMPORTS	393896	329099	252997	387054
of which				
Agricultural and food products	62763	51939	25041	32890
Food products and live animals	34824	31953	16908	22372
Fruit and vegetables	7238	7438	2364	4795

Source: Eurostat and EC Commission. Directorate-General for Agriculture. USA COMTRADE. The Agricultural Situation in the Community 1988 Report, Brussels 1989.

---

[2] The Agricultural Situation in the Community. 1987 Report, Brussels 1988.

To put world trade in fruit and vegetables into focus, in Table 2, the EEC exports and imports figures for these are set against those of the USA, for the relevant items; exports of all products, agricultural food products, live animals and finally fruit and vegetables. The USA is a larger net importer overall than the EEC but her agricultural exports are greater. The magnitude of fruit and vegetable exports is quite similar in both markets for 1986, but fruit and vegetable imports in the EEC were higher than the USA both in 1986 and 1980, and the rise of United States imports of fruit and vegetables in 1986 still left the EEC with much higher imports. This situation is associated with greater self-sufficiency in fruit and vegetables in the USA owing to the big range in climate, resulting in a large and diverse fruit and vegetable production, which represents an important intra-State trade. The notable increase in fruit and vegetable exports to the USA in recent years is associated with greater per capita consumption of staples and a bigger demand for exotic fruit and vegetables compatible with rising disposable incomes in the USA. Thus there has been an increase in the consumption of "winter fresh" fruit and vegetables from Central America and the Caribbean. The same trend is already under way in the EEC. It is likely that the EEC, much the biggest fruit and vegetable importer in the world, will continue to hold this position in future.

An Introduction to European Trade in Fruit and Vegetables

The amount of trade in fruit and vegetables is relative to a country's self-sufficiency in them. This is given in Table 3 for potatoes, other fresh vegetables, fruit other than citrus fruit and for citrus fruit. The source of this data is the official EEC annual report, The Agricultural Situation in the Community. Further analysis on self-sufficiency and other economic aspects of fruit and vegetables in the EEC follows in Chapter 2.

Table 3. EEC Self-sufficiency in Fruit and Vegetables 1985-6 - per cent

	Potatoes	Fresh Vegetables	Fruit Non-citrus	Citrus Fruit
EEC 12	102	107	87	75
EEC 10	103	101	83	47
Belgium-Luxembourg	109	118	63	-
Denmark	101	70	37	-
West Germany	92	38	55	-
Greece	105	150	130	163
Spain	99	133	114	328
France	102	91	89	4
Ireland	85	70	15	-
Italy	92	124	127	113
Netherlands	143	198	57	-
Portugal	88	146	95	100
United Kingdom	93	65	22	-

Source: Eurostat 1988. The Agricultural Situation in the Community, 1988 Report. Brussels 1989.

EEC Self-sufficiency in Vegetables

. Potatoes are the staple vegetable commodity throughout the Community. They are bulky and of comparatively low value such that international trade in potatoes is limited. The Community as a whole is self-sufficient in potatoes, though the degree of self-sufficiency in individual member states is varied. The Netherlands, for instance, is the largest exporter of potatoes in the Community and the Netherlands is also has an important potato starch industry. Lowest in the Community for self-sufficiency in potatoes is the Irish Republic, at 85 per cent. This comes as rather a surprise when one recalls the Irish potato famine of 1845 and its tragic consequences. Ireland is ideally suited to the production of potatoes. Countries like the UK and West Germany have a self-sufficiency in potatoes of between ninety and one hundred per cent. Such countries import the new season crop in the spring and early summer. Imports of new potatoes come from warmer climes such as Spain, Cyprus and Egypt. For instance, in the UK there is a sequence of sources reflecting differences in climate. The sequence starts with Egypt, then Cyprus and Spain, followed by the Channel Islands. Finally new potatoes come from the mild counties of Cornwall and Pembroke, which are all marketed before the large early potato crops in the East and South East of England. These new potatoes are followed by the English maincrop which starts its marketing season in August. New potatoes are more expensive than the later home maincrop, but consumers are willing to pay the price for their preference for new potatoes. Belgium-Luxembourg have a self-sufficiency in potatoes of 109 per cent, while Greece, France and Denmark are all rather more than self-sufficient. Portugal is only 88 per cent self-sufficient in potatoes. This suggests that Portugal at present has a less well developed fruit and vegetable sector than other member states in the EEC. It is quite likely, partly because of the recent political history of Portugal, that it will take Portugal at least a generation or so to catch up with her neighbour Spain.

The EEC overall is self-sufficient in vegetables, at 101 per cent for the 10 and 107 per cent for the 12. West Germany is conspicuously low in self-sufficiency at 38 per cent. The UK has a much higher self-sufficiency of 65 per cent. Denmark and Ireland are both 70 per cent self-sufficient and France rather more so, with a self-sufficiency of 91 per cent. The Netherlands heads the exporters of vegetables in the EEC by producing twice her domestic needs. Greece is also an exporter with a 150 per cent self-sufficiency in vegetables, followed by Portugal with a self-sufficiency of 146 per cent. Spain and Italy are also exporters with a self-sufficiency of 133 per cent and 124 per cent respectively. The figures in Table 3 do not disclose the high degree of competition between states in certain products, though the level of self-sufficiency does suggest the level of exports and imports between member states for the product group, vegetables. The EEC member states depend on each other to a high degree for a part of their vegetable supply.

EEC Self-sufficiency in Fruit - Other Than Citrus Fruit

For fresh fruit, other than citrus fruit, self-sufficiency for the 12 is only 87 per cent and for the 10, 83 per cent. Countries with a surplus for export are Greece, 130 per cent self-sufficient, Italy 127 per cent and Spain 114 per cent. However the UK and Ireland are both lacking in self-sufficiency in non-citrus fruit to a high degree. The UK is 22 per cent self-sufficient and Ireland is 15 per cent. Denmark is 37 per cent self-sufficient, West Germany is 55 per cent self-sufficient, and the Netherlands is 57 per cent self-sufficient. This sector in a given member state, depends heavily on intra-EEC imports and on third country imports. There is a wide range of fruit and vegetables in the retail shop or in the market place, and EEC consumers expect to have this range of choice.

EEC Self-sufficiency in Citrus Fruit

Although Spain is now in the EEC, and as a consequence raised the EEC self-sufficiency in citrus fruit from 47 per cent in the 10, self-sufficiency in citrus fruit in the 12 is still only 75 per cent. Spain is the great exporter of citrus fruit within the Community, with a self-sufficiency in citrus fruit of 328 per cent, or more than three times the amount consumed domestically. Spain dominates the citrus trade in the EEC. Greece, also a big exporter, has a self-sufficiency of 163 per cent, and Italy, by contrast, has a high home consumption. The smaller Italian exports of citrus fruit go mainly to West Germany. Italy has a self-sufficiency in citrus fruit of 113 per cent. Portugal is self-balancing, but all the northern states except France, which grows a nominal amount of citrus fruit, naturally, have no home supply of citrus fruit.

Community Exports and Imports of Fruit and Vegetables

Community exports and imports in thousand tonnes are given in Table 4 for 1985, 1986 and 1987. These export and import figures may be expected to vary with the size of the European crop in any given year.

Table 4. Community Exports and Imports by Product 1985-87 - '000 tonnes

	1985		1986		1987	
	Exports	Imports	Exports	Imports	Exports	Imports
Apples	189.2	512.1	172.7	577.2	186.1	584.6
Pears	58.1	101.7	53.7	105.9	149.8	154.4
Peaches	97.7	1.7	77.7	3.2	101.9	4.3
Oranges	328.0	862.0	447.7	914.1	393.4	891.8
Lemons	250.7	89.0	272.6	69.6	223.2	56.5
Tomatoes	71.6	271.9	75.9	264.7	88.6	273.8
Potatoes	610.5	319.1	612.6	380.5	827.9	343.4

Source: The Agricultural Situation in the Community 1988 Report Brussels 1989.

Exports of potatoes are some twice the quantity of imports. While imports of potatoes to the Community range from some 320 thousand tonnes to 380 thousand, exports vary from 610 thousand tonnes to 828 thousand tonnes over the three years. There is no particular trend and imports and exports of potatoes do vary with the EEC crop from year to year. Apple imports are three times, or nearly so, the quantity of apple exports in each of the three years. Imports lie between 512 and 585 thousand tonnes in the three years while exports are between 173 thousand and 189 thousand tonnes. Imports and exports of apples are seen to be comparatively stable over the three years, over which the European home crop has not fluctuated greatly. Pears show imports as 102 thousand tonnes in 1985, 106 thousand in 1986 and 154 thousand tonnes in 1987. Exports of pears were also large in 1987 and at 150 thousand tonnes they were nearly three times more than in the years 1986 and 1985. For peaches, imports from countries outside the EEC are negligible in each of the three years, while peach exports were 98 thousand tonnes in 1985, 78 thousand tonnes in 1986 and 102 thousand tonnes in 1987. This suggests that the EEC is more than self-sufficient for peaches with a regular export surplus.

Lemon imports lie between 57 and 89 thousand tonnes over the three years, with exports from 223 to 273 thousand tonnes. Exports of lemons well exceed imports. Orange exports range from 328 thousand tonnes to 448 thousand tonnes, and orange imports are much larger. They range from 862 thousand tonnes to 914 thousand tonnes. Tomato imports cover a narrow range of between 265 and 274 thousand tonnes.

The EEC Supply Balance for Vegetables

The EEC supply balance for vegetables from the Agricultural Situation in the Community annual report follows in Table 5. Further analysis of the trade balance for vegetables, for fruit except citrus fruit, and for citrus fruit, and for certain individual commodities from other reference sources will be given in Chapter 2.

Table 5. EEC Supply Balance, Vegetables 1986-7 - '000 tonnes

All Fresh Vegetables		Cauliflowers	
Production	46495	Sales by commercial producers	1648
Imports	650	Imports	
Exports	499	Exports	27
Intra-EEC	4084	Intra-EEC trade	287
Human consumption	38583	Human consumption	1493
Self-sufficiency per cent	107	kg per head	4.7
Kg per head	121.0		

All Fresh Tomatoes		Potatoes	
Sales by commercial producers	12622	Usable production	40852
Imports-fresh only	265	Imports	446
Exports-fresh only	76	Exports	962
Intra-EEC trade-fresh	788	Intra-EEC trade	5177
Human consumption	4201	Human consumption	25987
kg per head	13.4	Self-sufficiency per cent	101
		Kg per head	80.4

Source: Eurostat, The Agricultural Situation in the Community, 1988 Report, Brussels 1989.

The supply balance for vegetables, tomatoes, cauliflowers and potatoes is given in Table 5. Cauliflowers and tomatoes are included because data is available for these commodities since they are both régime commodities. The total vegetable production in the EEC is given in Table 5 as 46.5 million tonnes with imports of 650 thousand tonnes or only 1.4 per cent of home production, and exports of 499 thousand tonnes or only 1.1 per cent of home production. Intra - European trade in vegetables is 4084 thousand tonnes or 8.8 per cent of home production. There is some important intra-European trade in vegetables but it is clear that by much the most vegetables are marketed in the EEC member countries where the vegetables are grown. There is a balance in Table 5 of 38.6 million tonnes of vegetables for human consumption resulting in EEC consumption per head of 121 kg, from an EEC self-sufficiency in vegetables of 107 per cent.

Cauliflowers at 1.6 million tonnes represent only 3.5 per cent of all vegetables. In 1986-7 there were negligible exports and no imports. A relatively large proportion, 17.4 per cent, are traded within Europe. There are in fact several types of cauliflower according to the time in the year when the crop is harvested. In Britain we crop cauliflower for eleven months of the year. A proportion of the cauliflower crop put on the European market is winter cauliflower grown in the mild maritime climate in

Cornwall, Brittany and the Netherlands. Regular exports of cauliflower with special loading facilities are sent from Brittany to the UK over the period January to March when few substitute vegetables are available.

Table 6. EEC Supply Balance, Fresh Fruit 1986-7 - '000 tonnes

Fresh fruit, excl. citrus inc.preserves & juices

Usable production	19774
Imports	1081
Exports	578
Intra-EEC trade	3140
Human consumption	19212
Self-sufficiency per cent	88
Consumption per head kg	60.3

Apples

Sales by commercial producers	6569
Imports	491
Exports	173
Intra-EEC trade	1117
Human consumption	5945
Self-sufficiency per cent	88
Consumption per head kg	19.0

Pears

Sales by commercial producers	2241
Imports	102
Exports	53
Intra-EEC trade	234
Human consumption	2031
Consumption per head kg	6.3

Peaches

Sales by commercial producers	2870
Imports	3
Exports	78
Intra-EEC trade	
Human consumption	1558
Consumption per head kg	4.8

Grapes (1985-6)

Usable production	2606
Imports	589
Exports	140
Intra-EEC trade	77
Human consumption	2314
Consumption per head kg	7.6

Oranges (1985-6)

Usable production	4455
Imports	862
Exports	328
Intra-EEC trade	877
Human consumption	4548
Consumption per head kg	14.2

Citrus Fruit

Usable production	8000
Imports	1500
Exports	894
Intra-EEC trade	2226
Human consumption	9314
Self-sufficiency per cent	76
Consumption per head kg	29.5

Source: Eurostat, The Agricultural Situation in the Community, 1988 Report, Brussels 1989.

The supply balance for fresh tomatoes is in fact a misnomer since tomatoes for processing are included in sales by commercial producers. The figure for all tomatoes, 12.6 million tonnes is 27 per cent of all vegetables by volume. Imports of tomatoes are 265 thousand tonnes, which represent two per cent of home production of all tomatoes including those processed, and exports at 76 thousand tonnes represent only 0.6 per cent of EEC total tomato figure. Intra-European trade on the other hand is 788 thousand tonnes or 6.3 per cent of tomatoes for human consumption. When tomatoes for processing, at 7154 thousand tonnes, 98 thousand tonnes for livestock feed and losses of 1.2 million tonnes, in all 8.5 million tonnes, are deducted from the overall production of tomatoes of 12.6 million tonnes, this gives a figure of 4.1 million tonnes for human consumption. Intra-EEC trade in tomatoes is nineteen per cent of human consumption, exports are 1.8 per cent and imports 6.3 per cent of the figure for human consumption. Tomatoes are one of the most important commodities traded in European market. They are grown under glass with heat in the northern countries of the EEC and also in the open in the south with or without plastic protection and no heat at a different season.

The EEC Supply Balance for Fruit Other than Citrus Fruit

Referring to Table 6, the usable production of all fresh, except citrus, fruit is 19.8 million tonnes. Imports are 1081 thousand tonnes, exports 578 thousand and intra-EEC trade in fresh fruit is 3140 thousand tonnes. This results in a figure for human consumption of 19.2 million tonnes with consumption per head at 60.3 kg. Self-sufficiency in fresh fruit, other than citrus, in the EEC is 88 per cent.

Sales by commercial producers of pears is 2.2 million tonnes, imports are low at 102 thousand tonnes and exports are yet lower at 53 thousand tonnes. There is an intra-EEC trade in pears of 234 thousand tonnes and human consumption is 2 million tonnes. This gives a consumption per head of 6.3 kg, much lower than apples.

Apple sales by commercial producers is 6.6 million tonnes, imports are 491 thousand tonnes and exports are 173 thousand tonnes. There is a vigorous intra-EEC trade in apples of 1.1 million tonnes. This results in a figure for human consumption of 5.9 million tonnes and a consumption per head of 19 kg. Self-sufficiency in apple production in the EEC is 88 per cent. The usable production of grapes is 2.6 million tonnes but there is little intra-EEC trade. Imports of nearly 0.6 million tonnes, less lower exports of 140 thousand tonnes result in a human consumption figure of 2.3 million tonnes or 7.6 kg per head. Peaches show sales by commercial producers of 2.9 million tonnes, nominal imports and exports of 78 thousand tonnes. Losses to animal feed and marketing losses, together with processing and industrial use, leaves the figure for human consumption of peaches at 1558 thousand tonnes, which leads to a consumption per head of 4.8 kg for the EEC.

The EEC Supply Balance of Citrus Fruit

The usable production of citrus fruit is eight million tonnes. Intra-European trade is also large at 2.2 million tonnes as are imports at 1.5 million tonnes and exports of 0.9 million tonnes. This trade combined with a high EEC production, gives rise to a human consumption figure of 9.3 million tonnes or 29.5 kg per head. EEC self-sufficiency in citrus fruit is shown in Table 6 as 76 per cent. Oranges show a usable production figure of 4.5 million tonnes. The usable production of oranges is somewhat over half that of all citrus, and the sizable imports of 862 thousand tonnes are almost equal to the intra-EEC trade in oranges. Exports are 328 thousand tonnes. The effect of trade is to raise human consumption to 4.5 million tonnes, giving a consumption per head of 14.2 kg.

Some idea of the variability in fruit and vegetables from season to season is given in Table 7. For all vegetables and all fruit, data is given for the seasons 1984-5 to 1986-7 and data for individual commodities is given for the years 1985 to 1987. Exports and imports of vegetables as a whole show little difference between 1984-5 and 1985-6, the only two years shown. This applies to exports of fresh fruit but imports of fresh fruit were ten per cent higher in the first year than in the second. Over the three years exports of potatoes were similar at around 610 thousand tonnes in the first and second years, but exports of 828 thousand tonnes in 1986-1987 were 36 per cent up. Exports of tomatoes showed some change over the seasons, 72 thousand tonnes in 1984-1985, 76 thousand tonnes in

1985-1986 and 89 thousand in 1986-1987. Exports of apples also showed little variation from 189 thousand tonnes in the first year to 173 thousand in the second year and 186 thousand in 1987. Pear exports were more varied, at 58 thousand tonnes in 1985, 54 thousand in 1986 and 150 thousand in 1987. Pear imports were variable too, at 102 thousand in the first year, 106 thousand in the second and 154 thousand in 1987. Peach exports show considerable variance between the years. Exports are 98 thousand tonnes in 1985, 78 thousand in 1986 and 102 thousand tonnes in 1987. Imports of peaches are nominal over the three years. Trade in oranges shows great variability. Exports are 328 thousand tonnes for 1985, 448 thousand for 1986 and 393 thousand for 1987. Lemon exports show somewhat more stability. They are 251 thousand tonnes, 273 thousand tonnes and 223 thousand tonnes over the three years. Lemon imports were 89 thousand tonnes in 1985, 70 thousand in 1986 and 57 thousand in 1987.

Table 7. Community Trade in Fruit and Vegetables 1983-4 to 1986-7 - '000 tonnes

	1984-5		1985-6		1986-7	
	Exports	Imports	Exports	Imports	Exports	Imports
Fresh vegetables	**5165**	**2001**	**5120**	**2203**	**na**	**na**
Potatoes	610.5	319.1	612.6	380.2	827.9	343.4
Tomatoes	71.6	271.9	75.9	264.7	88.6	273.8
Fresh fruit	**1643**	**4818**	**1647**	**4378**	**na**	**na**
of which						
	1985		1986		1987	
Apples	189.2	512.1	172.7	577.2	186.1	584.6
Pears	58.1	101.7	53.7	105.9	149.8	154.4
Peaches	97.7	1.7	77.7	3.2	101.9	4.3
Oranges	328.0	862.0	447.7	914.1	393.4	891.8
Lemons	250.7	89.0	272.6	69.6	223.2	56.5

Source: The Agricultural Situation in the Community, 1988 Report, Brussels 1989.

Intra-Community trade in fruit and vegetables is shown in Table 8, for the seasons 1984-5 to 1986-7 in thousand tonnes. There is little change in incoming merchandise between 1984-5 and 1985-6 in the figure for fresh vegetables, but the three year figures given for tomatoes show a rise from 733 thousand tonnes in 1984-5 to 813 thousand in 1986-7. There is a similar change for potatoes which rise from 3.2 million tonnes in 1984-5 to 3.8 million tonnes in 1986-7. For 1984-5 and 1985-6 there is a change in incoming merchandise of fresh fruit in the first two years from 4.4 million to 4.7 million tonnes. Apples are relatively regular at some 1.2 million tonnes over the three seasons. Peaches are less regular at 415 thousand, 433 thousand and 508 thousand tonnes over the three seasons, showing a similar situation to oranges at 877 thousand tonnes, 1198 thousand tonnes and 1245 thousand tonnes over the three seasons. Incoming merchandise of lemons show an increase over the three seasons, 272 in the first, 302 thousand in the second and 331 thousand tonnes in the third.

Table 8. Intra-Community Trade in Fruit and Vegetables 1984-5, 1985-6 and 1986-7 - '000 tonnes

INCOMING MERCHANDISE	1984-5	1985-6	1986-7
Fresh vegetables	**6690**	**6999**	n.a
of which Tomatoes	733	788	813
Potatoes	3238	3377	3794
Fresh fruit	**4407**	**4730**	n.a
of which Apples	1200	1136	1358
Pears	280	237	307
Peaches	415	433	508
Oranges	877	1198	1245
Lemons	272	302	331
OUTGOING MERCHANDISE			
Fresh vegetables	**7697**	**8259**	n.a
of which Tomatoes	743	787	844
Potatoes	3109	3484	3857
Fresh fruit	**4740**	**5085**	n.a
of which Apples	1204	1133	1373
Pears	281	243	306
Peaches	420	417	497
Oranges	846	1427	1284
Lemons	291	322	344

Source: The Agricultural Situation in the Community, 1988 Report. Brussels 1989.

Outgoing merchandise of vegetables for the two years given in Table 8, varies between 7.7 million tonnes in the first season to 8.3 million tonnes in the second season. Tomatoes are 743 thousand tonnes in the first season, 787 thousand in the second season and yet higher in the third, at 844 thousand tonnes. Potatoes also show considerable change, 3.1 million tonnes in the first season, 3.5 million in the second and 3.9 million tonnes in 1986-7. Outgoing merchandise of fresh fruit shows little difference in the two seasons given, and apples are relatively regular. Pears are 281 thousand tonnes in the first season, 243 thousand in the second and over 300 tonnes in the third. Peach exports show the amount in the first and second seasons to be almost the same at some 420 thousand tonnes to rise to almost 500 thousand tonnes in 1986-7. The orange trade is most variable. In the first season, outgoing merchandise of oranges is 846 thousand tonnes, in the second 1427 thousand, and in the third 1284 thousand tonnes. Lemons, too, are variable in the outgoing trade, from 291 thousand tonnes in the first season, to 322 in the second and 344 thousand tonnes in the third season.

Intra-EEC trade in fruit and vegetables, for 1980 and for the four consecutive years 1984 to 1987, is given in Table 9. The data comes from Eurostat, The Agricultural Situation in the EEC, and it is only for the EEC 10 as no similar data is available for the 12. The point to remember is that Spain, the great trader in fruit and vegetables, is not included. Also all the data given in Table 9 relates to régime commodities which trade is monitored in some detail for régime procedure to be carried out. Fruit and vegetables are termed régime commodities when, apart from attracting other features of the

EEC fruit and vegetable trade policy, they are given an intervention system or more protection by the application of the reference price system to them. The data in Table 9 gives intra-EEC trade, imports and exports for the three classes, vegetables, fruit other than citrus fruit, and for citrus fruit.

Intra-EEC trade in vegetables as a whole is seen to be on a rising trend, from nearly three million tonnes in 1980 to over four million tonnes in 1987. Intra-EEC trade in cauliflowers is quite variable from year to year if not erratic, but over the years the general level of the cauliflower trade has shown little change. Intra-EEC trade in tomatoes, much the most important vegetable in this trade, rose from 549 thousand tonnes in 1980 to 813 thousand in 1987. A similar development took place for aubergine, which is a minor vegetable, but since relatively recently aubergine was raised to the rank of a régime commodity, aubergine trade data is available as a by-product of the régime procedure. Tomatoes feature much less in the exports, but exports of tomatoes are on a fairly steeply rising trend, from 38 thousand tonnes in 1980 to 89 thousand in 1987. A comparatively small amount of cauliflowers of some 25 to 31 thousand tonnes are exported but scarcely any cauliflowers are imported. Imports of aubergines run at between eight and thirteen thousand tonnes, but there are small exports of from two to three thousand tonnes.

Intra-EEC trade in non-citrus fruit is quite variable, but it would seem to be on a rising trend, taking one year with another, being 2.7 million tonnes in 1980 and 3.6 million tonnes in 1987. Apples represent the largest single item in the trade, at 1.1 million tonnes in 1980 to 1.3 million tonnes in 1987. Peaches are next in importance, and are seen to rise over the years from 0.35 million to 0.5 million tonnes. Pears come next and the high EEC trade in pears rose from 257 thousand tonnes in 1980 to just over 300 thousand tonnes in 1987. Apricots show a low intra-EEC trade of some forty to sixty thousand tonnes.

Imports of non-citrus fruit range around the million tonne mark over the years to be substantially above this in 1987. Exports of this produce, which are more than half the imports, are on a rising trend, more strongly than that for imports. From 1980 to 1987 non-citrus fruit exports rose from 478 thousand tonnes to 735 thousand tonnes. Apples and pears are by far the most important non-citrus fruit imports, together they comprise around half the non-citrus fruit imports. In 1980 Pear imports amounted to one fifth of the apple imports but pear imports have risen in recent years. In 1980 pear imports were 76 thousand tonnes, but by 1987 pear imports have doubled. At this time pear imports had become 28 per cent of the apple and pear group. Apple imports are on a lower, but sustained rising trend, from 401 thousand tonnes in 1980 to 543 thousand tonnes in 1987. Of the other non-citrus fruit only peaches and apricots are listed. Imports of both these fruits are small. Apple exports are appreciably lower than apple imports.

Imports of citrus fruit, between 1.7 and 1.5 million tonnes, have been quite irregular over the years, but in 1986 and 1987 they were 1.5 million tonnes in each year, of which some 60 per cent is oranges. Imports of lemons are sustained at a lower level which shows marked deviations from year to year. Exports of citrus fruit are increasing such that the exports in 1987 were more than three-fold those in 1980. Exports of citrus fruit were 0.2 million tonnes in 1980, but from 1984 to 1987 citrus fruit exports were over 0.6 million tonnes and in 1986 they reached 0.9 million tonnes. Overall, orange imports, though very irregular from year to year are fairly static over the years with a possible tendency to drop. Imports of lemons are also very irregular from year to year with a range over the years of below 45 thousand tonnes to 89 thousand. Exports of lemons in contrast have more than doubled since 1980, a trend which may be expected to continue.

Imports of vegetables are modest at some six to seven hundred thousand tonnes and tomatoes are the most important of these. Imports of non-citrus fruit are up and down over the years but their average level is some eleven thousand tonnes. Apples are the most prominent non-citrus fruit in this trade, followed by pears. The import of citrus fruit shows no particular trend over the years. Oranges are the most important commodity shown in imports of citrus fruit. Vegetable exports are relatively minor in quantity and they vary quite considerably from year to year. Most of the vegetables exported are vegetables other than those specified individually in Table 9. The prominent exports are apples, pears and peaches. Citrus fruit exports are again larger than those of vegetables and both oranges and lemons are important exports.

Table 9. Intra-EEC Trade in Fruit and Vegetables 1980-87 - '000 tonnes

		1980	1984	1985	1986	1987
Vegetables		**2977**	**3622**	**3801**	**4084**	**4135**
of which	Cauliflowers	224	276	190	287	237
	Tomatoes	549	692	733	788	813
	Aubergines	25	32	32	35	36
Fruit excluding citrus		**2700**	**2990**	**3168**	**3140**	**3635**
of which	Apples	1078	1145	1163	1117	1293
	Pears	257	267	276	234	302
	Peaches	346	425	415	433	507
	Apricots	45	53	51	61	58
Citrus Fruit		**1784**	**2305**	**1872**	**2226**	**2404**
of which	Oranges	923	1190	877	1198	1245
	Lemons	253	299	272	301	331
IMPORTS						
Vegetables		**603**	**613**	**662**	**650**	**730**
of which	Cauliflowers	-	-	1	-	1
	Tomatoes	236	262	272	265	274
	Aubergines	12	10	13	8	9
Fruit excluding citrus		**981**	**1002**	**1014**	**1081**	**1210**
of which	Apples	401	445	461	491	543
	Pears	76	78	101	102	151
	Peaches	3	2	2	3	4
	Apricots	6	4	6	4	3
Citrus Fruit		**1722**	**1259**	**1419**	**1520**	**1529**
of which	Oranges	1073	761	862	914	862
	Lemons	78	45	89	70	57
EXPORTS						
Vegetables		**341**	**436**	**457**	**499**	**593**
of which	Cauliflowers	31	29	21	27	25
	Tomatoes	38	64	72	76	89
	Aubergines	2	2	2	3	3
Fruit excluding citrus		**478**	**576**	**622**	**578**	**735**
of which	Apples	204	189	189	173	184
	Pears	60	62	58	53	149
	Peaches	58	80	98	78	102
	Apricots	5	16	18	15	18
Citrus Fruit		**203**	**606**	**685**	**894**	**752**
of which	Oranges	80	247	328	448	393
	Lemons	115	243	251	273	223

Source: Eurostat. The Agricultural Situation in the Community, 1988 Report, Brussels 1989.

Table 10. EEC Trade With Mediterranean Countries, ACP Countries, and Member States' Overseas Territories 1980, 1986 and 1987 - million ECUs

	IMPORTS			EXPORTS			TRADE BALANCE		
	1980	1986	1987	1980	1986	1987	1980	1986	1987
MEDITERRANEAN TRADE									
All Products	21350	23277	23981	23855	30185	28226	2505	6908	2454
Agricultural products	2120	2816	2804	3192	3449	3209	1073	633	405
Fruit and vegetables	1228	1696	1661	166	186	178	-1062	-1510	-1483
ACP and MEMBER STATES OVERSEAS TERRITORIES TRADE									
All products	21092	20225	16871	17432	17403	15376	-3660	-2822	-1495
Agricultural products	6250	9016	7641	2365	2575	2256	-3885	-6441	-5385
Fruit and vegetables	336	625	604	113	150	141	-233	-475	-463

Source: Eurostat, EEC Commission Directorate General for Agriculture: The Agricultural Situation in the Community 1988 Report. Brussels 1989

The EEC Fruit and Vegetable Trade with Mediterranean Countries, and Other Trade Groupings

The Community has an important trade in fruit and vegetables with non-EEC Mediterranean countries. Table 10 gives imports and exports and the trade balance for 1980, 1986 and 1987 for all products, agricultural products, and for fruit and vegetables. For all products EEC exports exceed imports to leave a modest but sustained trade balance and the same applies to exports and imports of agricultural products. When it comes to fruit and vegetables, however, imports are seen to be between 1.2 and 1.7 billion ECUs and exports are under 0.2 billion ECU. (The European Unit of Account has a value which varies from some 60 to 65 pence sterling.) Trade with non-EEC Mediterranean countries includes Cyprus, Israel, the Magreb and Mashraq countries and Turkey, most of whom have been granted concessions on the EEC tariffs for fruit and vegetables by the EEC over a number of years. The ACP countries, most of which come with the category of least developed developing countries, are chiefly former French possessions in Africa, and a number of Caribbean, African and Pacific countries, all of whom qualify for concessions on the EEC tariff for fruit and vegetables. ACP and member states' overseas territories include a number of suppliers of exotic fruit and vegetables for which there is a growing demand in Europe. Table 10 shows a trade deficit for all three trade categories, all agricultural products and for fruit and vegetables. The level of imports of fruit and vegetables from the Mediterranean countries in 1986 and 1987 are nearly three times the volume of fruit and vegetable imports from ACP countries and member states' overseas territories, but the imports from these countries is rising fast and it is taking an important part in their steps towards economic development. It is arising from 336 million ECUs in 1984 to 604 million ECUs in 1987, an increase of 86 per cent, whereas fruit and vegetable imports from the non-EEC Mediterranean countries increased by only a third between 1980 and 1987.

CHAPTER 2

THE EEC FRUIT AND VEGETABLE SECTOR THE SUPPLY BALANCE OF FRUIT AND VEGETABLES

Production, Trade and Consumption

The chief determinant of the level and composition of trade in fruit and vegetables of any country or trading block such as the European Economic Community, is the amount and the nature of its own fruit and vegetable supply. There is a range of self-sufficiency between fruit and vegetable commodities so that the amount of produce below self-sufficiency needs to be imported. This accounts for the considerable intra-EEC trade. Also a large part of European supply is grown in the hotter climatic belts over the world. This features in the important external trade of the EEC. It is this situation of partial self-sufficiency or lack of it, which gives rise to the character and composition of both intra-EEC and extra-EEC trade.

Climate and Diversity of Production

Production of fruit and vegetables in the EEC is diverse and this great diversity is occasioned by differences in climate. Starting from the northern limits of the EEC it is seen that Britain, for instance, lies between the latitude 50° north (the latitude cuts the English channel) and latitude 60° north (the latitude of the Shetland Islands). All countries in this belt have a north temperate climate and for several countries like Britain and Ireland, Holland and regions of France the climate is modified by the sea to become north temperate maritime, a milder form of the north temperate climate. This particular type of climate gives rise to the production of a number of winter vegetables. At the same time in England, the Wash in Lincolnshire marks the northern limit for commercial apple and pear production. Countries and regions in Europe, away from the sea, have a continental climate with severe winters, a short spring and autumn, and hotter summers. Such conditions account for the large Polish export of onions which benefits from the hot summer for ripening.

The Mediterranean climate enjoyed by Spain, Portugal, the south coastal region of France, Italy and Greece is characteristically mild and warm, and because of the near absence of frost is given to a large production of citrus fruit much of which features in intra-EEC trade. The latitude 40° north cuts through the middle of Spain, Italy and Greece. Before Spain joined the Community in 1986, Community self-sufficiency for citrus was some 40 per cent; with Spain in the market, the EEC is now over 70 per cent self-sufficient for citrus fruit. It is the warmer summer and longer hours of sunshine than further north, which gives rise to the pre-eminence of France in the apple market of the EEC. These conditions lead to high yields and high quality.

Figure 1. The Fruit and Vegetable Sector in the EEC

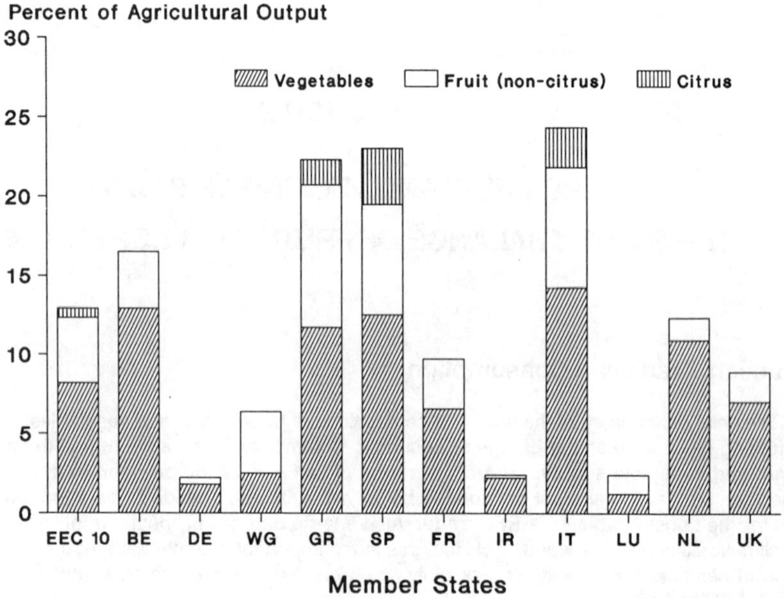

Per cent of agricultural output.

	Vegetables	Fruit (non-citrus)	Citrus
EEC 10	8.2	4.1	0.6
Belgium	12.9	3.6	-
Denmark	1.8	0.4	-
West Germany	2.5	3.9	-
Greece	11.7	9.0	1.6
Spain	12.5	7.0	3.5
France	6.6	3.1	-
Ireland	2.2	0.2	-
Italy	14.2	7.6	2.5
Luxembourg	1.2	1.2	-
Netherlands	10.9	1.4	-
UK	7.0	1.9	-

Source: Eurostat 1989

Figure 2. The Balance of Trade in the EEC, 1986-7, Vegetables

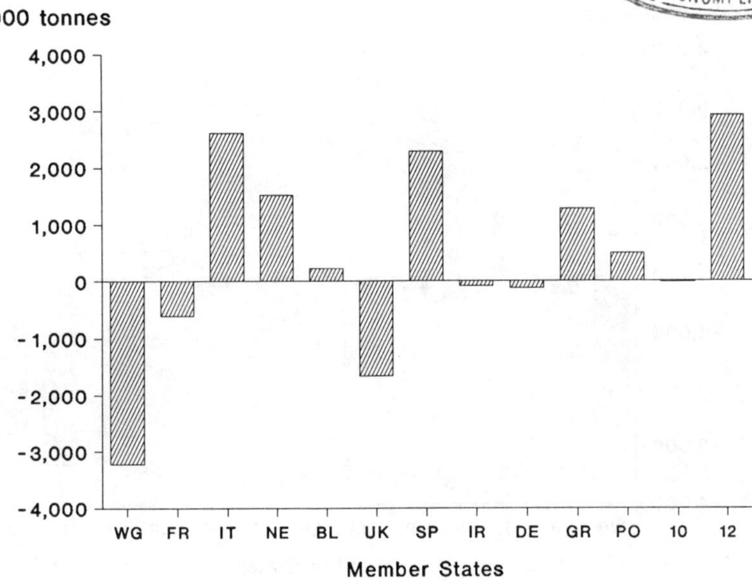

Trade Balance

	'000 tonnes
West Germany	-3225
France	-608
Italy	2608
Netherlands	1518
Belgium/Luxembourg	228
UK	-1658
Spain	2281
Ireland	-87
Denmark	-123
Greece	1280
Portugal	494
EEC 10	-18
EEC 12	2917

Source: ZMP Bilanz Gemüse 1987-8

Figure 3. The Balance of Trade in the EEC, 1986-7
Fruit (excluding citrus)

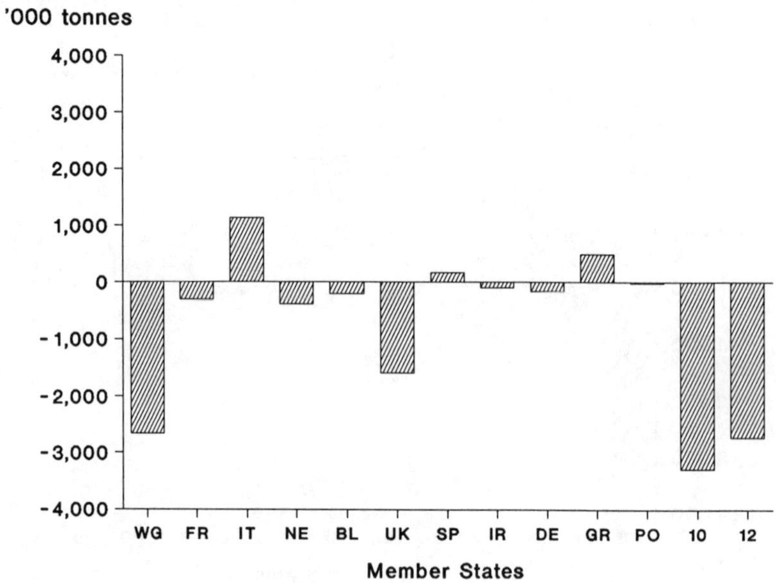

Trade Balance

	'000 tonnes
West Germany	-2662
France	-305
Italy	1138
Netherlands	-381
Belgium/Luxembourg	-209
UK	-1586
Spain	176
Ireland	-100
Denmark	-160
Greece	493
Portugal	-13
EEC 10	-3284
EEC 12	-2731

Source: ZMP Bilanz Obst 1987-8

Figure 4. The Balance of Trade in the EEC, 1986-7
Citrus Fruit

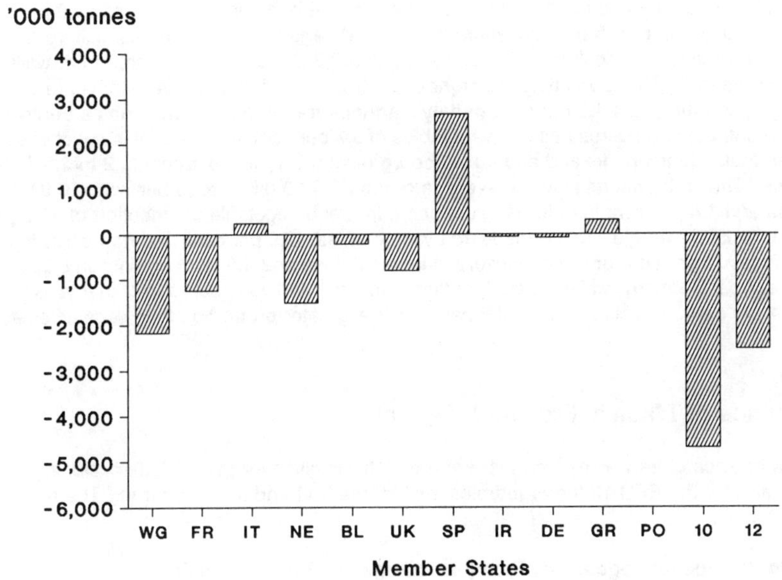

Trade Balance

	'000 tonnes
West Germany	-2169
France	-1250
Italy	243
Netherlands	-1513
Belgium/Luxembourg	-209
UK	-804
Spain	2657
Ireland	-58
Denmark	-73
Greece	314
Portugal	-1
EEC 10	-4687
EEC 12	-2521

Source: ZMP Bilanz Obst 1987-8

The Share of Fruit and Vegetables in Agricultural Output

The proportion of the agricultural output, which is defined as the total agricultural product by value, is illustrated for non-citrus fruit, citrus fruit and for vegetables. Figure 1 shows the proportion by value of fruit and vegetables in each member state, of the total value of each country's agricultural output. Data on Portugal is not available. Portugal apart, only Greece, Spain and Italy grow citrus fruit. The amount in France is negligible. These three countries on account of their Mediterranean climate have the highest proportion of agriculture in fruit and vegetables. Of the total output in Greece, 11.7 per cent by value is in vegetables, nine per cent is in non-citrus fruit and 1.6 per cent is in citrus fruit, making up a 22.3 per cent share for fruit and vegetables of Greek agriculture as a whole. In Italy 14.2 per cent of agriculture is in vegetables and 7.6 per cent in non-citrus fruit, while 2.5 per cent is in citrus fruit. This gives Italy the highest contribution of fruit and vegetables of all member countries, accounting for 24.3 per cent of Italy's agricultural output. France has a contribution to agricultural output in non-citrus fruit and vegetables of 9.7 per cent, 6.6 per cent in vegetables and 3.1 per cent in fruit. Denmark, Ireland and Luxembourg have a very low per cent, 2.2 to 2.5, in fruit and vegetables. The Netherlands is at the average of the EEC 10 taken together, with 10.9 per cent in vegetables and 1.4 per cent in fruit. Belgium has a fruit and vegetable contribution of 16.5 per cent, more than the EEC average. This is made up by vegetables, 12.9 per cent, and non-citrus fruit 3.7 per cent. In Britain 7 per cent of the agriculture is in vegetables and 1.9 per cent in non-citrus fruit. By contrast, in West Germany, while the contribution of the fruit and vegetable sector in total is on the low side at 6.4 per cent, non-citrus fruit at 3.9 per cent is a greater proportion than vegetables at 2.5 per cent.

The Balance of Trade in Fruit and Vegetables

The balance of trade is exports less imports. This is given for the EEC 10, excluding Spain and Portugal, and for the EEC 12 for vegetables, non-citrus fruit and for citrus fruit. The figures are in thousand tonnes.

The Trade Balance for Vegetables, Fruit (excluding citrus) and Citrus Fruit

The balance of trade for vegetables is given in Figure 2. The inclusion of Spain and Portugal results in the EEC 12 having a positive vegetable trade balance of 2917 thousand tonnes, as opposed to an almost imperceptible negative balance for the EEC 10 of 18 thousand tonnes. The countries with a positive trade balance for vegetables, apart from Spain and Portugal, are Greece, Italy and Holland. It has been seen that Spain, Italy and Greece have a large contribution of fruit and vegetables in their agriculture. The country with the largest positive vegetable trade balance is Italy, at 2608 thousand tonnes. Next comes Spain, 2281 thousand tonnes, followed by Holland with a very high net vegetable balance for her size and climate of 1518 thousand tonnes. West Germany stands out as having the biggest negative balance by a large margin of 3225 thousand tonnes. Next to West Germany, the country with a large negative vegetable trade balance of 1658 thousand tonnes, is Britain. France has a small vegetable trade deficit of 608 thousand tonnes. Greece has a positive vegetable trade balance of 1280 thousand tonnes, and Portugal, a positive vegetable trade balance of 494 thousand tonnes. Belgium achieves a small net vegetable trade balance of 228 thousand tonnes and Ireland and Denmark both have a small negative balance. Ireland and Denmark are both small countries with a relatively small population, upwards of three million in Ireland and five million in Denmark.

The balance of trade for fruit excluding citrus is illustrated in Figure 3. The country with the largest positive balance is Italy, with a balance in non-citrus fruit of 1138 thousand tonnes. The only other country with a positive trade balance of non-citrus fruit is Greece with 493 thousand tonnes. The EEC 10 has a negative trade balance of non-citrus fruit of 3284 thousand tonnes and the EEC 12 has one of 2731 thousand tonnes. West Germany has much the largest negative non-citrus fruit balance, of 2262 thousand tonnes, followed by Britain with a negative fruit excluding citrus trade balance of 1586 thousand tonnes. The Netherlands, France, Belgium-Luxembourg, Ireland and Greece all have small negative non-citrus fruit trade balances of around 200 to 300 thousand tonnes.

The balance of trade in citrus fruit is featured in Figure 4. As might be expected, Spain, with a large citrus fruit production, also has a large positive citrus fruit balance of 2657 thousand tonnes. Italy, Spain and Greece, all with a large production, have positive trade balances for citrus fruit. However, it should not be forgotten that data for Portugal is not available. All other countries in the EEC, except Portugal, do not produce any citrus fruit, which needs a Mediterranean climate. The country with the highest imports of citrus fruit and the largest negative trade balance is West Germany, Greece has a small positive balance of 314 thousand tonnes and Italy has a yet smaller balance of 243 thousand tonnes. It is clear that domestic consumption of citrus fruit in Greece and Italy is high as they export only a small part of their production. West Germany is the biggest market in the EEC for citrus fruit, with a negative balance of 2169 thousand tonnes. Holland has a negative trade balance for citrus fruit of 1513 thousand tonnes. Apart from Britain, with a negative trade balance of 804 thousand tonnes, the negative trade balance for citrus fruit in the other countries is small, because of their small population, and therefore their citrus fruit imports are also small. The overall effect is a large negative citrus fruit trade balance for the EEC 10 of 4687 thousand tonnes, which is reduced for the EEC 12 to a negative citrus fruit balance of 2521 thousand tonnes.

Self-sufficiency and Consumption in the EEC

This section deals with self-sufficiency and consumption in member states and over the EEC 12 as a whole and over the EEC 10. It is in the fruit and vegetable sector of the EEC that big changes followed the accession of Spain and Portugal in 1986. It must be emphasised that, more particularly, the changes were brought about by Spain, a country already advanced in the export of fruit and vegetables. Spain is a big source of citrus fruit, accounting traditionally for example, for some sixth of the entire import of fruit and vegetables to Britain. It is the sheer size of the Spanish supply and her potential to supply more exports which makes Spain so important. Spanish producers have the technical knowledge to excel in production and, coupled with this, Spain has developed the necessary infrastructure for the transport and export of fruit and vegetables. Even in the mid-term, production and exports of a wide range of fruit and vegetables will increase as the transition period for Spain entering the EEC progressively comes to an end. The ten year transition will end in 1996.

Trade in fruit and vegetables is in a major part determined by the relative self-sufficiency of production in member states and by the propensity to consume in member states. Different member states have different self-sufficiencies and any one country needs to trade to make up for deficiencies in home demand for fruit and vegetables. In practice a given country will turn first to supplementing its requirements from partners in the Common Market which they share. Again this policy may not meet the level of requirement for certain produce, and exotic produce is necessarily purchased from countries outside the Community.

A high rate of self-sufficiency leading to a surplus is the policy of some countries, often for a range of products. The outstanding example is Holland, which over the years was soon established as the market garden for Germany before World War II and for West Germany since. Through sheer technical prowess, efficiency and competitiveness, Holland sought to be, and has become, an essential part of the market garden of Britain. Holland's strength lies in vegetables, but Holland like Belgium, successfully exports apples and pears. Belgium's part in the trade is overshadowed by Holland, but in a smaller measure the Belgium production of fruit and vegetables and her trade in this produce is notable in its own right. It is of note that, in Belgium, the contribution of fruit and vegetables to agricultural output is higher than that in Holland.

Vegetables

Self-sufficiency and consumption of vegetables in the EEC and in member countries is illustrated in Figure 5. The EEC 12 as a whole have a self-sufficiency for fruit and vegetables of 106.9 and in the EEC 10 one of 99.9 per cent which suggests a self-balancing market. This market includes exports to countries like Scandinavia and Switzerland. Despite the apparently high self-sufficiency of the EEC for vegetables, there is a considerable amount of intra-EEC trade. In particular, produce is imported at certain times of the year according to consumer preference which itself varies between member states. This is affected greatly by tradition, income levels and the composition of supply. For the EEC 12 the consumption of vegetables is high at an average of 119.7 kg per head and a little less

Figure 5. Self-Sufficiency and Consumption in the EEC 1986-87, Vegetables

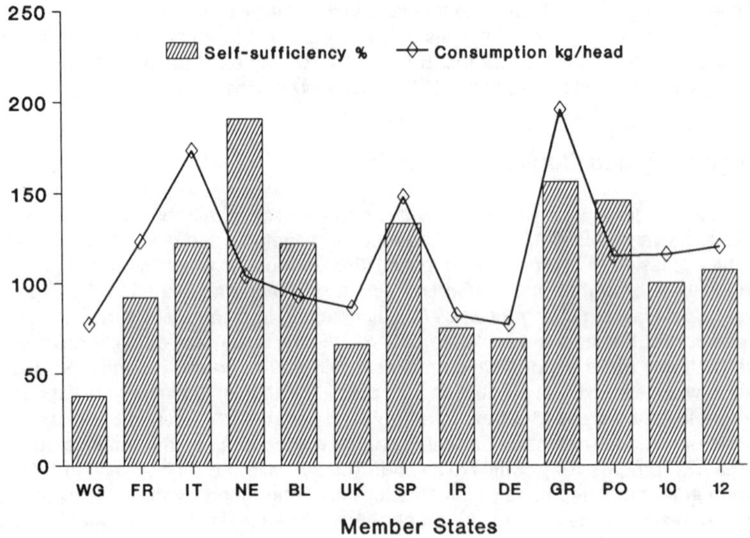

Self-Sufficiency and Consumption

	Self-sufficiency per cent	Consumption per head (kg)
West Germany	37.8	77.4
France	92.1	123.0
Italy	122.3	173.3
Netherlands	190.8	103.9
Belgium/Luxembourg	121.9	92.9
UK	66.2	86.2
Spain	132.8	147.9
Ireland	75.1	82.2
Denmark	68.9	77.1
Greece	155.8	195.4
Portugal	145.5	114.8
EEC 10	99.9	115.6
EEC 12	106.9	119.7

Source: ZMP Bilanz Gemüse, 1987-8

Figure 6. Self-Sufficiency and Consumption in the EEC 1986-87, Fruit (excluding citrus)

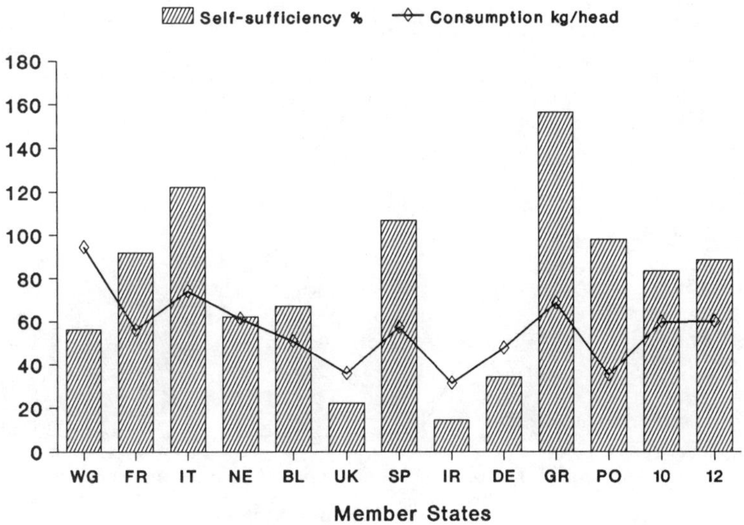

Self-Sufficiency and Consumption

	Self-sufficiency per cent	Consumption per head (kg)
West Germany	56.1	94.4
France	91.7	56.0
Italy	122.1	74.0
Netherlands	62.0	61.0
Belgium/Luxembourg	66.9	50.6
UK	22.2	35.8
Spain	106.6	57.0
Ireland	14.5	31.3
Denmark	34.2	47.4
Greece	156.1	68.3
Portugal	97.6	35.1
EEC 10	83.0	59.4
EEC 12	88.2	59.6

Source: ZMP Bilanz Obst, 1987-8

Figure 7. Self-Sufficiency and Consumption in the EEC 1986-87, Citrus Fruit

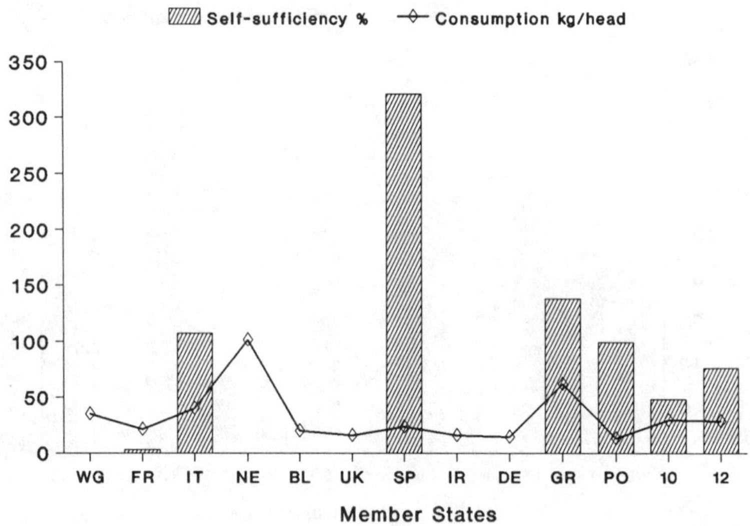

Self-Sufficiency and Consumption

	Self-sufficiency per cent	Consumption per head (kg)
West Germany		34.8
France	2.9	21.4
Italy	107.2	39.9
Netherlands		101.5
Belgium/Luxembourg		20.0
UK		15.9
Spain	320.7	23.4
Ireland		15.8
Denmark		14.2
Greece	137.7	62.0
Portugal	99.3	12.8
EEC 10	48.1	29.6
EEC 12	76.0	28.6

Source: ZMP Bilanz Obst, 1987-8

for the EEC 10 at 115.6 kg per head. Between member states self-sufficiency and consumption vary greatly. Of course, to a certain extent, the degree of self-sufficiency in any one member state determines the amount of supplies brought in. Each country has its own preferences. Perhaps obvious examples are the consumption of sauerkraut in West Germany and of chicory in Belgium.

West Germany is by a long way lowest in self-sufficiency for vegetables in the EEC. Next door is the Netherlands with a self-sufficiency of 190.8 which is in line with the very large Dutch exports of vegetables. The consumption of vegetables in Holland is modest at 103.9 kg per head, below average for the EEC as a whole. This demonstrates that the Dutch vegetable economy has clearly been developed to provide a surplus for export, much the larger share going to West Germany. The West German vegetable economy is consequently small because of the large Dutch exports. Other countries with over one hundred per cent self-sufficiency in vegetables, and some vigour in the export trade, are Greece at 155.8 per cent, Portugal at 145.5 per cent, Spain at 132.8 per cent and Belgium-Luxembourg at 121.9 per cent. Domestic consumption in Greece is much the highest of all member states at 195.4 kg per head. Next is Italy at 173.3 kg per head, then Spain, 147.9 kg. Vegetable consumption in Portugal is 115 kg per head. Greece, Italy and Spain all have high domestic consumption and high exports. Portugal has a consumption only at the average of the EEC 10, and low exports. Other countries of low self-sufficiency are Britain 66.2 per cent, Denmark 68.9 per cent and Ireland 75.1 per cent. Consumption per head of vegetables in Denmark is the lowest in the EEC at 77.1 kg per head, but the other countries with low self-sufficiency have low consumption also. In Ireland, the per capita consumption of vegetables is 82.2 kg, and in the United Kingdom it is little higher at 86.2 kg per head, much below the average for the EEC 12 of 119.7 kg per head.

Fruit excluding Citrus Fruit

Self-sufficiency and consumption of fruit (excluding citrus fruit) is illustrated in Figure 6. Countries more than self-sufficient are Greece at 156.1 per cent, Italy at 122.1 per cent and Spain at 106.6 per cent. These countries, known for their production of citrus fruit, are also the largest producers of non-citrus fruit. The EEC 12 is self-sufficient for this class of fruit at the level of 88.2 per cent and the EEC 10 is much less self-sufficient at 83 per cent. Consumption per head for the EEC 12 is 59.6 kg and is less for the EEC 10. By far the largest consumer of non-citrus fruit is West Germany where the consumption per head is 94.4 kg and self-sufficiency a low 56.1 per cent, which means a big reliance by West Germany on imports. Italy with a high self-sufficiency is next to West Germany in consumption per head, but at 74.0 kg per head this is considerably below that of West Germany. Greece has a consumption per head of non-citrus fruit of 68.3 kg. Spain, in the export league, has a comparably low consumption of 57 kg per head. France has a similar per head consumption and is over 90 per cent self-sufficient in this type of fruit. While Portugal is almost self-sufficient, consumption per head is next to the lowest in Europe at 35.1 kg. Lowest in self-sufficiency in non-citrus fruit is Ireland, but the United Kingdom is also very low. This very low degree of self-sufficiency in the UK has come about by a policy, originally protecting this sector and reinforced in 1947, being necessarily very substantially changed when the UK became a member of the EEC in 1973.

Citrus Fruit

Self-sufficiency and consumption of citrus fruit in the EEC is illustrated in Figure 7. The outstanding country for citrus fruit is Spain. It has a 320 per cent self-sufficiency, most of which is geared to the export market, with a modest consumption of 23.4 kg per head to match this situation. Italy with a self-sufficiency of 107.2 per cent has an above average consumption of citrus fruit of 39.9 kg per head. Greece with a self-sufficiency of 137.7 per cent has a very high consumption of 62 kg per head. France has a small production on the Mediterranean leading to a self-sufficiency of only three per cent. French consumption of citrus fruit is slightly below the average for the EEC 12. West Germany has a higher than average consumption of citrus fruit of 34.8 kg per head. Outstanding is the consumption of citrus fruit in the Netherlands of 101.5 kg per head. In stark contrast, its neighbour, Denmark, has the lowest consumption of 14.2 kg per head. Consumption per head of citrus fruit in Britain also low at 15.9 kg per head, with Irish consumption at 15.8 kg per head.

Table 1. EEC Balance Sheet 1986-7 '000 tonnes

(Where 1986-87 figures not available, 1985-6 is given in parenthesis)

	12	10	WG	FR	IT	NE	BL	UK	SP	IR	DE	PO	GR
Vegetables													
Useable production	{46495}	{35063}	1959	{7048}	14289	3189	1269	3244	9226	263	272	{1900}	3575
Imports	{2203}	{3465}	3453	{1557}	340	844	651	2156	86	113	171	{3}	23
Total supply	{48698}	{38528}	5412	{8605}	14629	4033	1920	5400	9312	376	443	{1903}	3598
Exports	{5120}	{3447}	228	{949}	2948	2362	879	498	2367	26	48	{497}	1303
Change in stocks	{100}											{100}	
Total domestic use	{43478}	{35081}	5184	{7656}	11681	1671	1041	4902	6945	350	395	{1306}	2295
Losses	{4556}	{3419}	453	{850}	1580	36	60		1001	58		{105}	344
Animal feed	{730}	{436}		{5}	160	117	30		212			{32}	
Human consumption	{38192}	{31226}	4731	{6801}	9941	1518	951	4902	5732	292	395	{1169}	1951
Self sufficiency %	{106.7}	{99.9}	37.8	{92.1}	122.3	190.8	121.9	66.2	132.8	75.1	68.9	{137.8}	155.8
Consumption/head/yr kg	{119.7}	{115.6}	77.4	{123.0}	173.3	103.9	92.9	86.2	147.9	82.2	77.1	{114.8}	195.4
Fruit (excluding citrus)													
Useable production	{19774}	{15643}	3393	{3385}	6294	622	423	452	2857	17	83	{380}	2135
Imports	{4378}	{4861}	3854	{1322}	730	1018	585	1680	437	116	207	{17}	8
Total supply	{24152}	{20504}	7247	{4707}	7024	1640	1008	2132	3294	133	290	{397}	2143
Exports	{1647}	{1577}	1192	{1017}	1888	637	376	94	613	16	47	{4}	501
Total domestic use	{22505}	{18927}	6055	{3690}	5136	1003	632	2038	2681	117	243	{393}	1642
Losses	{1714}	{1222}	279	{510}	380	78	59		363	5		{20}	197
Animal feed	{214}	{143}			98	38	30		71				42
Human consumption	{20577}	{17562}	5776	{3180}	4658	887	543	2038	2247	112	243	{373}	763
Self sufficiency %	{88.0}	{83.0}	56.1	{91.7}	122.1	62.0	66.9	22.2	106.6	14.5	34.2	{96.7}	130.02
Consumption/head/yr kg	{59.5}	{59.2}	94.4	{56.0}	74.0	61.0	50.6	35.8	57.0	31.3	47.4	{35.1}	76.5
Citrus Fruit													
Useable production	{8000}	{4350}		{37}	3614				3861			{136}	1146
Imports	{4051}	{5566}	3107	{1307}	142	2608	434	834	22	62	93	{1}	14
Total supply	{12051}	{9916}	3107	{1344}	3756	2608	434	834	3883	62	93	{137}	1160
Exports	{1530}	{879}	938	{57}	385	1095	225	30	2679	4	20		328
Total domestic use	{10521}	{9037}	2169	{1287}	3371	1513	209	804	1204	58	73	{137}	832
Losses	{1148}	{870}	41	{105}	997	30	4		294	2		{7}	212
Industrial use	{20}	{20}			20				3				
Human consumption	{9353}	{8147}	2128	{1182}	2354	1483	205	804	907	56	73	{130}	620
Self sufficiency %	{76.0}	{48.1}		{2.9}	107.2				320.7			{99.3}	137.7
Consumption/head/yr kg	{28.9}	{29.7}	34.8	{21.4}	39.9	101.5	20.0	15.9	23.4	15.8	14.2	{12.8}	62.0

Source: Eurostat 1989

The EEC Supply Balance for Fruit and Vegetables - Production, Trade and Consumption

The balance sheet for vegetables, non-citrus fruit and citrus fruit for member states and for the 10 and 12, is given in Table 1. These come from Eurostat data for 1986-7 and where gaps occur 1985-6 data is used.

The Supply Balance for Vegetables

Starting with usable home production at the top of Table 1, total supply to each member state or group is obtained by the addition of the quantity of vegetables imported. From the figure of total supply, exports are deducted as they diminish the supply of the commodity concerned. The large exporters of the EEC with vegetable exports of between two and three million tonnes are Italy, with virtually three million, Spain with 2.4 million and the Netherlands, also with 2.4 million, Greece exports 1303 thousand tonnes of vegetables and Belgium-Luxembourg 879 thousand tonnes. Countries with exports of 500 or less are the UK, Portugal, West Germany and Ireland.

Usable home production of vegetables in the 12 in Table 1 is seen to be 46 million tonnes, and that for the 10, 35 million tonnes. The large increase of usable home production when the Community enlarged to 12 member states in 1986, is very substantially due to usable production in Spain which greatly increased the EEC total. Italy has the largest production of vegetables with 14 million tonnes, followed by Spain with 9.2 million tonnes. France comes next with a usable production of 7 million tonnes. Other countries in the EEC have a much smaller usable production of vegetables. For instance, Greece has only 3.5 million tonnes while the UK has a usable production of 3.2 million tonnes. West Germany, for its size, has a very small production of vegetables of 2 million tonnes, much less than the UK for instance, while Belgium-Luxembourg has a large production for its size of 1.3 million tonnes. Denmark and Ireland each have a small production of between two and three hundred thousand tonnes. A change of stocks is present for the 12 only and this is a low figure. There follows in Table 1 the total amount for domestic use. This is much the highest for Italy which leads from her large domestic use figure of 11.7 million tonnes. Spain, like France, comes near Italy with a usable production of 6.9 million tonnes. Italy has the largest exports of all member states, 2.9 million tonnes, and much the highest human consumption of vegetables of 9.9 million tonnes. (The figure for human consumption is derived from that for domestic use net of animal feed and losses, where applicable). Domestic use is high in France at 7.7 million tonnes. This level of domestic use reflects the high usable production of vegetables in France, of seven million tonnes and the excess of imports over exports. Imports are 1.6 million tonnes.

Usable production in Spain at 9.2 million tonnes is higher than that in France. For Spain this leads to a total domestic use of seven million tonnes which remains high after the deduction of large exports of vegetables of 2.4 million tonnes. Substantial losses of one million tonnes are shown for Spain which might be expected of a country with such high exports of quality produce. Also, as one might expect Spanish imports of vegetables are negligible at 86 thousand tonnes.

West Germany shows a small usable production of vegetables for her size, of two million tonnes. Imports by West Germany raise total supplies in West Germany to 5.4 million tonnes with human consumption at 4.7 million tonnes, a little less than that in the UK. The usable production of vegetables in the UK is 3.2 million tonnes and imports are 2.2 million, giving a total supply of 5.4 million tonnes. Exports are small at 0.5 million tonnes and domestic use in the UK is 4.9 million tonnes.

Usable production in the Netherlands is 3.2 million tonnes and exports of 2.4 million tonnes lead to a domestic use of 1.7 million tonnes, high for the population in the Netherlands. Domestic use in Belgium-Luxembourg at 1.0 million tonnes is also high for the size of the population. Denmark and Ireland, both small countries with small populations, show a similar domestic use of 395 and 350 thousand tonnes respectively. From the total domestic use, losses and animal feed is deducted to give the figure for human consumption, one of the principal aims of Table 1. In the 12, losses are some ten per cent and animal feed is small to give a human consumption of 38.2 million tonnes. The 10 have human consumption at the lower figure of 31.2 million tonnes with losses in the same proportion of the 12, and again a low amount for animal feed. Allocation of animal feed from usable production is only featured in six countries. It is highest in Spain at 212 thousand tonnes. Animal feed is also high in Italy at 160 thousand tonnes. In the Netherlands the figure for animal feed is again high at 117 thousand tonnes, but losses are nominal. In these three countries, all of which grow and export a large amount of vegetables, the discard rate, which builds up the figure for animal feed and losses may be expected to be high because these exporters can only afford to put quality produce on the export market. Italy's losses are also high, at 1.6 million tonnes, which is relative to a degree to Italy's exceedingly high level of usable product and also to the business of exporting. Losses are comparatively high in France, though at half the level in Italy. Losses in West Germany are 0.5 million tonnes. This reflects the high quality standards of the trade in West Germany. Losses in Greece are 344 thousand tonnes and in Belgium-Luxembourg, the Netherlands and Ireland, they are as low as 60 thousand tonnes or less.

Self-sufficiency in vegetables on Eurostat figures for member states is variable as one would expect. Self-sufficiency in vegetables is much the highest in the Netherlands at 191 per cent. It is next highest in Greece, at 156 per cent, but it is high also in Spain at 133 per cent and in Italy at 122 per cent. From this level of self-sufficiency is reflected the high level of vegetable exports, between 2 and 3 million tonnes, in the case of Italy, Spain, and the Netherlands, and 1.3 million tonnes in Greece.

Consumption per head, again from Eurostat data, is like self-sufficiency, quite variable between member states. For the 10, vegetable consumption per head is 116 kg and for the 12 it is higher at 120 kg. The country with much the highest per head consumption of vegetables in Greece at 195 kg. Next is Italy at 173 kg, then Spain at 148 kg. Related to the high consumption in Italy and Greece is the lower share of their high usable production given to the export market. Consumption and the size of home production do not always relate, although often they do. West Germany with self-sufficiency of only 38 per cent and with a low usable production, has a per head consumption of vegetables of 77 kg. Belgium-Luxembourg with a self sufficiency of 122 per cent and a relatively high usable production of 1.3 million tonnes has a per head consumption of vegetables of 93 kg. The UK with a self-sufficiency of 66 per cent and consumption of 86 kg per head is well below the EEC average. However, all countries which are self-sufficient with capacity for exporting vegetables except the Netherlands and Portugal have a consumption per head well above the average for the EEC 12 of 120 kg. In Greece the role of the vegetable sector is seen to be substantially to feed the Greek population with exports taking a quite proper secondary role. Consumption in the Netherlands where exports are 30 per cent more than the amount for human consumption, the per head vegetable consumption is relatively modest at 104 kg. Looking at Italy, Greece and Spain where self-sufficiency in vegetables is 122 per cent for Italy, 156 per cent for Greece and 133 per cent for Spain, vegetable consumption per head in kg is 173, 195 and 148 respectively. Italy's exports of vegetables are virtually 3 million tonnes, those of Greece 1.3 million tonnes and those of Spain 2.4 million tonnes. France has a high consumption of 123 kg per head which is associated with imports of 1.5 million tonnes and a high usable production of 7 million tonnes. Per capita consumption in France is also, of course, like that in any member state, associated with consumer preferences and tastes. These in turn are associated with national tradition and custom.

The Supply Balance for Fruit excluding Citrus

Usable production of non-citrus fruit given in Table 1 is substantially apples and pears. This is nearly 20 million tonnes for the 12 and for the 10, 15.6 million tonnes. The addition of Spain to the 10 increased the EEC non-citrus fruit sector by some three million tonnes. The figures for the 10 and the 12 cannot be reconciled as a number of figures are presented using 1985-6 data in the absence of data for 1986-7. Italy has the largest usable production of non-citrus fruit at 6.3 million tonnes.

West Germany and France vie for second place, each having a non-citrus fruit usable fruit production of 3.4 million tonnes, but so far as apples and pears are concerned the West German production does not match the French for dessert quality. Spain is next with a usable production of 2.9 million tonnes. The Netherlands and Belgium-Luxembourg each have a small orchard area, which is substantially apples and pears intensively grown on modern systems and well balanced for varieties in demand. The Netherlands usable production of non-citrus fruit is 0.6 million tonnes and in Belgium-Luxembourg it is some 0.4 million tonnes. Usable production in the UK is in size between that of the Netherlands and Belgium-Luxembourg at 0.45 million tonnes. With a smaller usable production comes Portugal with 0.38 million tonnes.

Imports of non-citrus fruit are affected by the level of the harvest in each country in each season, higher home production meaning fewer imports. This fluctuates from year to year in the member states mainly depending on the distribution, timing and severity of spring frosts. In a given year damaging frost can be severe throughout Western Europe, or it may only be critical in certain regions, and sometimes there is a maritime-continental divide, which by reducing yield in one part of Western Europe gives rise to more than otherwise intra-European trade, when the production in this situation in one EEC sector is less than that in another. A good proportion of non-citrus fruit is grown in a temperate climate. West Germany has outstandingly the largest imports at 3.8 million tonnes, to be followed by the UK at 1.7 million tonnes and France, a big producer and exporter at 1.3 million tonnes. The Netherlands, a relatively large producer and exporter, imports one million tonnes, and Belgium-Luxembourg takes about half this amount or 0.6 million tonnes. Denmark imports 0.2 million tonnes of non-citrus fruit but Ireland, with very little production, imports only a small amount, 133 thousand tonnes. Portugal, as one might expect has nominal imports, but also what may come as a surprise, nominal exports. Portugal, because of her history since 1945 is not a force in international or intra-EEC trade in fruit and vegetables. It will take a generation at least for Portugal to feature prominently in international trade in fruit and vegetables. Her only strength is in the sub-sector of tomato paste and this commodity is in surplus in the EEC. Greece has a usable production of non-citrus fruit of over 2 million tonnes. As one would expect her imports are minimal giving Greece a total supply of much the same figure. Greek exports are 0.5 million tonnes and losses, animal feed and processing account for the large figure of nearly 0.9 million tonnes which results in that for human consumption of 0.8 million tonnes. This leads to a high consumption per head in Greece of 76.5 kg. Consumption per head for the 10 and the 12 is 59 kg. Greece has a self-sufficiency of 130 per cent which relates to the policy of Greece of high home consumption.

As mentioned earlier, in the EEC 12, usable production of non-citrus fruit is 2.0 million tonnes and in the 10, 15.6 million tonnes. Imports to the 12 are 4.4 million tonnes and in the 10 are 4.9 million tonnes. Total supply is 24.1 million tonnes for the 12 and 20.5 million tonnes for the 10. Exports of non-citrus fruit for the 12 and the 10 are very similar at 1.6 million tonnes, since almost all the production of non-citrus fruit, is in the 10. West Germany, with a usable production of non-citrus fruit of 3.4 million tonnes, imports more than this amount, 3.9 million tonnes. The result is that the figure for human consumption in West Germany at 5.8 million tonnes, is higher than that of any other member state. This gives a high consumption per head for non-citrus fruit in West Germany of 94.4 kg. West German self-sufficiency in non-citrus fruit is 56 per cent. France has a smaller usable production of non-citrus fruit of 3.4 million tonnes and France imports 1.3 million tonnes. France exports quite a considerable amount of non-citrus fruit, one million tonnes, after losses are taken into account, total domestic use in France of non-citrus fruit is 3.7 million tonnes resulting in a figure for human consumption of 3.1 million tonnes. The consumption per head of non-citrus fruit in France is 56 kg, but her self-sufficiency in this sector is notably high at 91.7 per cent.

Italy has a usable production of 6.3 million tonnes and exports of 1.9 million tonnes - more than double her imports. Total domestic use is high at 5.2 million tonnes but human consumption is much lower at 4.2 million tonnes, the difference being losses and animal feed. Consumption per head in Italy of non-citrus fruit is well below that of West Germany at 74 kg per head, but only slightly less than Greece, 76.5 kg per head. The Netherlands has a small usable production of non-citrus fruit of 0.6 million tonnes. Imports amount to 1.0 million tonnes with exports of 0.6 million tonnes, leaving one million tonnes for domestic use. Consumption of non-citrus fruit in The Netherlands is nearly 1 million tonnes and consumption per head 61 kg. Self-sufficiency in this sector is 62 per cent.

Belgium-Luxembourg have a domestic use of 0.63 million tonnes and imports of 0.6 million tonnes with exports of 0.37 million tonnes, just over half that of The Netherlands. From a total domestic use of 0.63 million tonnes, human consumption per head in Belgium-Luxembourg is 50.6 kg and self sufficiency is 66.9 per cent, a little above Holland.

The UK has a usable production of non-citrus fruit of 0.45 million tonnes. The UK imports are high at 1.7 million tonnes to give a total UK supply of 2.1 million tonnes. Exports from the UK are very low so that total domestic use is 2.0 million tonnes which corresponds with the figure for human consumption. UK self-sufficiency in this sector is low at 22.2 per cent. As referred to elsewhere this low figure is substantially due to successful competition from France in the UK dessert apple sector.

Spain has a usable production of non-citrus fruit of 2.9 million tonnes. Imports are small at 0.4 million tonnes which gives a total supply in Spain of 3.3 million tonnes. Exports are only modest at 0.6 million tonnes but Spain has the potential to export considerably more non-citrus fruit. Spain's total domestic use is 2.7 million tonnes and the figure for human consumption is 2.2 million tonnes. The

Table 2. EEC Vegetable Production 1987, Areas and Quantities

Items as listed in Eurostat

	'000 ha		'000 tonnes	
	EEC 12	EEC 10	EEC 12	EEC 10
Cauliflowers	126.7	114.4	1960.6	1705.5
Brussels sprouts	24.1	-	336.7	-
Cabbage	55.0	47.0	1785.8	1564.5
Other brassicas	46.4	-	1109.3	-
All brassicas	**252.2**	**161.4**	**5192.4**	**3270.0**
Celeriac and celery	15.7	13.6	535.2	480.3
Leeks	27.1	24.1	669.1	607.2
Lettuce	80.7	53.3	1973.9	1319.5
Asparagus	54.1	32.2	195.3	113.0
Chicory	45.8	45.5	571.4	568.0
Globe artichoke	88.9	63.6	856.1	518.0
Other, leafed or stalked	36.5	31.1	825.3	709.9
Vegetables leafed or stalked	**348.8**	**263.4**	**5626.3**	**4315.9**
Tomatoes	242.9	167.1	11250.6	8133.4
Cucumbers	15.4	9.3	1172.1	838.5
Gherkins	7.9	6.1	160.2	139.4
Melons	112.2	42.6	1718.1	789.6
Water melons	67.1	39.4	1980.1	1350.7
Aubergines	19.7	15.3	539.6	413.6
Pumpkins, marrows	28.2	21.1	795.0	565.0
Peppers	50.1	22.2	1291.8	584.7
Other cultivated for fruit	2.2	2.2	18.1	18.1
Cultivated for fruit	**545.7**	**325.3**	**18925.6**	**12833.0**
Kohl-rabi	1.6	1.6	44.0	44.0
Turnips	13.7	12.2	372.4	348.8
Carrots	64.2	55.0	2314.1	2053.9
Garlic	57.4	14.9	362.1	123.8
Onions	99.7	66.4	2989.0	1813.9
Shallots	2.8	2.8	38.9	38.9
Beetroot	9.1	8.5	305.3	291.6
Salsify and scorzonera	3.8	3.8	69.0	69.0
Radish	6.2	5.3	149.3	135.2
Other roots and tubers	6.6	4.5	112.1	79.3
Roots and tubers	**265.1**	**175.0**	**6756.2**	**4998.4**
Peas	144.7	134.3	562.8	540.0
French beans and runner beans	120.6	91.3	960.0	667.3
Other pulses	55.4	37.3	442.7	282.9
Pulses	**320.7**	**262.9**	**1965.5**	**1490.2**
Cultivated mushrooms	-	-	**599.3**	**551.2**
Other fresh vegetables	**81.7**	-	**1167.0**	-
Total of vegetables specified	**1814.2**	**1188.0**	**40232.3**	**27458.7**

Source: Eurostat 1989

Spanish self-sufficiency in this sector is 106.6 per cent and her consumption per head is 57 kg, a figure 2 kg below the average in EEC 12. It is likely that Spain's exports of non-citrus fruit will expand in future, particularly apples and pears, at the expense of France.

Ireland has scarcely any usable production of non-citrus fruit. Her imports amount to only 116 thousand tonnes and with small exports she has 117 thousand tonnes available for domestic use. Irish self-sufficiency in non-citrus fruit is a very low, 14.5 per cent and her consumption per head is the lowest in the EEC at 31.3 kg. While Ireland's potential for growing non-citrus fruit is not great, her self-sufficiency in this sector is lower than it could be.

The Supply Balance for Citrus Fruit

Only four countries grow sufficient citrus fruit to show up in the Eurostat statistics presented in Table 1 but there is known to be a very little production in France. Portugal is seen to have 136 thousand tonnes of citrus fruit most of which is consumed in Portugal. Italy has a usable production of 3.5 million tonnes, Spain has a larger usable production of 3.9 million tonnes and Greece one of 1.1 million tonnes. Italy, Spain and Greece import negligible quantities of citrus fruit. Italy exports 0.4 million tonnes and has a total domestic use of 3.4 million tonnes. In Italy losses are nearly one million tonnes, leaving for human consumption only 2.3 million tonnes. Self-sufficiency in Italy is 107.2 per cent and consumption is virtually 40 kg per head, the highest of all member states but Greece.

It is quite evident that Italy's policy towards citrus fruit is to maintain home consumption at a high level which means smaller exports of citrus. Spain by comparison with a usable production of 3.9 million tonnes, exports as much as 2.7 million tonnes leaving a low amount for domestic use of 1.2 million tonnes. Losses are relatively high so that the amount for human consumption is only 0.9 million tonnes. The very high figure for self-sufficiency of 320.7 per cent together with the low consumption per head, indicates Spain's policy which is to export as much as possible. Consumption per head in Spain is only 23.4 kg per head. Italian consumption per head is 70 per cent higher than that in Spain, but consumption in Portugal at 12.8 kg per head is nearly half that of Spain. Italy with a consumption per head of 39.9 kg, is second only to Greece and the Netherlands in Europe. Spain with a usable production of citrus of 3.9 million tonnes exports 2.7 million tonnes, leaving a low domestic use of 1.2 million tonnes and a figure for human consumption of only 0.9 million. There is a very high self-sufficiency in citrus in Spain, which accounts for her large exports. Greece has a usable production of citrus of 1.1 million tonnes and a human consumption, after exports and losses, of 0.7 million tonnes. This results in consumption per head of 62 kg, second only to the Netherlands in the EEC. Greece, one of the larger exporters retains more citrus fruit for home consumption than she retains of other fruit and of vegetables. The Greek self-sufficiency for citrus fruit is 137.7 per cent.

West Germany imports 3.1 million tonnes of citrus, of which 0.9 million tonnes is exports, to satisfy an above average consumption of 34.8 kg per head. France imports one million tonnes and has a consumption per head of 21.4 kg. The Netherlands imports 2.6 million tonnes and exports one million giving the Netherlands a very high consumption per head of 101.5 kg. Belgium-Luxembourg also trade in citrus with exports over half her imports of 434 million tonnes. The exports are entrepôt trade. The UK trades less, and with a figure for human consumption of citrus of 0.9 million tonnes. Her consumption per head is 15.9 kg. Consumption per head in Ireland is the same as that in the UK and that in Denmark is a little lower. One point of importance is that Spanish entry to the Community raised the self-sufficiency of the EEC 10 from 48.1 to 76.0 per cent in the 12.

The Production and Area of Fruit and Vegetables in the EEC

The area and production of vegetables in the EEC is given in Table 2 and that of fruit in Table 3. Data for the 10 is shown against that for the 12 to illustrate the impact of Spain and Portugal. Portugal alone has little impact, partly on account of her small size, but mainly because of the limited development related to her recent political régime.

Table 3. EEC Fruit Production 1987, Areas and Quantities

Items as listed in Eurostat

	'000 ha		'000 tonnes	
	EEC 12	EEC 10	EEC 12	EEC 10
Fresh grapes	**258.1**	**184.0**	**3017.8**	**2437.4**
Apples	339.2	259.6	8042.6	6912.2
Pears	132.0	87.2	2605.4	2039.3
Apples and pears	**471.2**	**346.8**	**10648.0**	**8951.5**
Peaches	211.8	140.2	2706.8	2133.6
Apricots	62.2	37.2	545.8	397.2
Cherries	86.9	61.1	629.8	552.5
Plums	62.0	40.0	865.8	723.4
Nectarines	-	40.1	537.4	466.6
Other stone fruit	1.0	0.9	10.5	10.5
Stone fruit	**423.9**	**319.5**	**5296.1**	**4283.8**
Black and red currants	7.8	7.8	166.7	166.7
Strawberries	50.8	39.4	701.2	467.5
Raspberries	5.8	5.8	59.2	59.1
Gooseberries	1.4	1.4	88.6	88.6
Kiwi	21.3	20.1	150.1	149.3
Other soft fruit	23.8	12.0	201.5	120.5
Soft fruit	**110.9**	**86.5**	**1367.3**	**1051.7**
Oranges	299.7	141.9	4467.1	1924.4
Mandarins	30.5	20.5	304.3	238.3
Satsumas	15.9	-	485.5	3.8
Clementines	63.4	19.5	963.3	193.8
Lemons	110.5	52.4	1467.3	710.9
Grapefruit	1.8	0.7	27.3	8.2
Other citrus fruit	5.7	3.2	61.4	44.9
Citrus fruit	**527.5**	**238.2**	**7776.2**	**3124.3**
Figs	129.5	23.4	178.0	87.2
Quinces	2.9	0.5	28.4	11.7
Avocados	6.2	0.3	33.2	0.7
Other fruit NOS	30.0	12.2	564.8	21.1
Other fruit	**168.6**	**36.4**	**804.4**	**120.7**
Total of fruit specified	**1960.2**	**1211.4**	**28909.8**	**19969.4**

Source: Eurostat 1989

The Vegetable Sector in the EEC

Vegetables, of which there is great variety, account for 1.8 million ha, 2.7 per cent, out of a total arable area in the EEC of 67.8 million ha. The great range of vegetable crops is occasioned by the diverse climates in Western Europe, from north temperate maritime to north temperate, both with relatively mild winter seasons, to the type of continental climate in Central Europe, which is characterised by extremes of temperature between summer and winter, to the Mediterranean climate with mild wet winters and warm dry summers.

The classification of vegetables in Table 2 shows that of the total vegetable area in the EEC 12, brassicae crops cover fourteen per cent of the total area of vegetables of which half is cauliflower, and the leafed and stalked group of vegetables, which include the important salad vegetables, cover twenty per cent of the entire vegetable area of EEC 12. Vegetables cultivated for fruit, an EEC term which is botanically correct, cover thirty per cent of the total vegetable area. Forty-five per cent by area of the vegetables in this class is tomatoes which also includes tomatoes for processing. Roots and tubers cover fifteen per cent of the entire EEC vegetable area and thirty eight per cent of the area of the group roots and tubers is accounted for by onions. Onions are an example of a crop which can be grown successfully in the northern countries of the EEC in the temperate zone but they can be grown (different varieties) most successfully and to a higher quality standard in the Mediterranean region, particularly in Spain where production has reached one million tonnes. Pulse crops account for eighteen per cent of the EEC 12 vegetable area.

Enlargement of the EEC to include Spain and Portugal increased the total vegetable area from 12 to 18 million ha, an increase of 53 per cent. The largest change was in the group of vegetables cultivated for fruit of 68 per cent. This group includes a number of important Mediterranean vegetable crops. For instance, the area of peppers and melons cultivated more than doubled. Water melons increased in area by seventy per cent. The area of cucumbers increased by 66 per cent and that of tomatoes by 45 per cent and the area of roots and tubers increased by 51 per cent. The garlic crop increased nearly four fold and onions increased in area by fifty per cent. The area of leafed and stalked vegetables increased by 32 per cent,including asparagus 68 per cent, lettuce 51 per cent and chicory forty per cent. The area under pulse crops increased by 22 per cent with the biggest change taking place in the unspecified pulse crops.

Table 2 gives the production volume of vegetables for the 10 and the 12 in thousand tonnes. With total production at forty million tonnes, almost half or 18.9 million tonnes is the group of vegetables cultivated for fruit, of which tomatoes alone are over one million tonnes. The more bulky roots and tubers amount to 17 per cent of the total volume of vegetables and again the bulky brassicae amount to 13 million per cent. Leafed and stalked vegetables are 5.6 million tonnes or 14 per cent of the total volume of vegetables. Lettuce alone is nearly two million tonnes. Pulse crops are five per cent of the total EEC volume of vegetables. These crops are of relatively high value.

The differences in the figures of crop area and crop volume of the 10 and the 12 of course, illustrate the impact of EEC enlargement in the fruit and vegetable sector. Overall, while the area of vegetables cultivated increased by 53 per cent, production increased by only 47 per cent. For brassicae crops the increase in the production by volume is greater than the increase in area because of high yields in Spain, while for leafed and stalked vegetables the reverse holds. The asparagus crop increased in production by 73 per cent, globe artichokes by 65 per cent and lettuce by fifty per cent. Vegetables cultivated for fruit had an increase in area of 68 per cent but an increase in production of 47 per cent. Roots and tubers show an increase in area of 51 per cent and an increased in production of 35 per cent. There is some reason to explain the high yields of the brassicae, in that inclusion of these commodities from a Mediterranean climate would include higher yielding crops, but there is no rationale to explain why vegetables should have a relatively lower production. It may be that overall yields were low in the 1987 season.

The Fruit Sector of the EEC

In Table 3 it is seen that the EEC fruit sector of the 12 covers virtually two million ha or 2.9 per cent of the entire EEC arable area. Thirteen per cent of the area is accounted for by table grapes and almost a quarter by apples and pears. The group of stone fruit is nearly equal in area to that of apples and pears at 22 per cent. Soft fruit covers only six per cent of the fruit area but strawberries account for 46 per cent of the area under all soft fruit. The biggest share of the fruit area in the 12 is citrus fruit, which covers 27 per cent of the entire fruit area of the 12. Oranges alone account for 57 per cent of the considerable area down to citrus fruit in the EEC.

Enlargement of the EEC increased the area of fruit in the Community by 62 per cent. While fruit production in the Community increased by 45 per cent on EEC enlargement from 10 to 12, the increase in the EEC apples and pears area is one of 36 per cent with an increase of 19 per cent in apple and pear production by volume. The apple portion of the apple and pear group is 72 per cent by area and 76 per cent by volume of production. It is likely that apples and pears in the Iberian peninsula do not have the optimum climatic conditions for production as do those countries further north which means that the Iberian apples may have comparatively lower yields. However, Iberian apples do have the advantage of an earlier harvest by two or three weeks as compared with the French.

The area of stone fruit increased on enlargement by 33 per cent and in volume by 24 per cent. The area of peaches, which account for precisely half the area and half the production of stone fruit, increased in production on enlargement by 27 per cent. Apricots, 15 per cent of the area of stone fruit, increased in production by 37 per cent while plums, 26 per cent of the stone fruit area, increased in production by 20 per cent. Soft fruit, on enlargement, increased in area by 28 per cent and in production by 30 per cent. Strawberries, the crop largest in size among the soft fruit, increased in production by fifty per cent.

The citrus fruit sector was more than doubled while production increased by 150 per cent. Oranges comprised 57 per cent of the citrus area of the EEC 12 and they increased in production by 132 per cent. Lemons, 21 per cent of the citrus area, increased in production by 106 per cent. Thus the need was met for consumers in the EEC to have ready access to citrus fruit. Of the fruit in the residual category in Table 3 which amounts to 8.6 per cent of the fruit area, over three quarters of this area is under figs. The area of figs in EEC 12 increased more than five-fold and production increased nearly three times. The change is partly accounted for by the highly favourable climate for figs in the south of Spain. The area and production of land down to fruit in the EEC, some two million ha producing thirty million tonnes of fruit is given in Table 3.

The same diversity of climate puts fruit into distinct crop sectors. For example, apples and pears tolerate winter frost and actually benefit from winter dormancy, while citrus fruit require a Mediterranean climate with little or no frost in the winter, which is the same requirement for much of the stone fruit. Fruit production is more limited by climate than vegetables, since this group of produce comprises all true fruits botanically, temperature is more critical to bud initiation, and the number of hours of sunshine determines fruit maturity, fruit quality and yields. Also, fruit, unlike vegetables, by its very nature is more labour intensive, with higher labour requirements for pruning and harvesting. Fruit, compared with most vegetables, has a special appeal to consumers which gives rise to a high popular demand and a preparedness on the part of the consumer to buy it.

Apples and pears, 471 thousand ha, normally at much lower prices than the out-of-season imports, account for 24 per cent of the fruit area and 36 per cent EEC fruit production. Citrus fruit, with 527.5 thousand ha, accounts for 27 per cent of the EEC fruit area and 27 per cent of the production. Then there are 423.9 thousand ha of stone fruit, or 22 per cent of the entire area of fruit in the EEC and 18 per cent of the production. Thus, three quarters of the EEC fruit area and 85 per cent of the production is cropped by perennial fruit trees. These have a commercial life span of from twenty to thirty years with perhaps up to five years before the trees begin to bear. This represents high committed investment, the more so with intensive systems which are now being practised. There is a much smaller area of soft fruit, 110.9 thousand ha, less than 6 per cent of the fruit area, and also a comparatively large area of figs, 129.5 ha.

Intra-European trade in fruit is also carried out to a high degree. This is occasioned by the distinct and natural climatic regions required for the production of different types of fruit. Again fruit, as compared with vegetables is generally of high value whereas this only applies to a much smaller proportion of vegetable production. Citrus fruit, of which the largest quantities are oranges and lemons, account for virtually a quarter of the total fruit tonnage or 7.8 million tonnes. A considerable amount of citrus fruit is concerned in intra-European trade to markets in the north where the climate is too cold for growing citrus. Citrus fruit is, of course, grown in the Mediterranean region which has optimum climatic conditions for growing quality citrus fruit. Stone fruit, 5.3 thousand tonnes or 18 per cent of all EEC fruit production by volume, comprises peaches, nectarines, apricots plums and cherries. Production of these fruits in the Mediterranean region occasions a considerable amount of intra-

European trade because the Mediterranean fruits are on the market up to two months before produce in the same season grown in the north temperate countries. The early strawberry production from the Mediterranean zone is also traded in this way. Some strawberries are exported to Europe, from Kenya and Columbia for instance, over our winter season. In the off-season of European supply of fruit, there is an important export trade to Europe from countries in the Southern hemisphere. This situation means that the European consumer is well supplied with the various kinds of fruit throughout the whole year.

Economic Aspects of Fruit and Vegetable Production

Fruit and vegetables are characterised by the need for intensive culture and their perishability. At the same time a majority of vegetables are produced under protected conditions but the actual harvesting of vegetable crops, in particular of salad vegetables, still requires hand operation.

While most vegetables are annual crops, most fruit crops are perennial, which has implications for investment. Fruit crops require by and large much more capital and much more labour than annual crops. Also, like many vegetable crops, fruit production is intensive in the use of land. Apples and pears for instance have an economic life-span of from twenty to thirty years with perhaps up to five years before the trees begin to bear. This represents a high commitment of investment, the more so with intensive systems which are now being practised. The high annual labour requirement in top fruit production is inescapable because of the demands for pruning and harvesting. Much of the soft fruit, notably strawberries, is cropped for three or four years, while cane fruit may have an economic life of from ten to twelve years. Both these types of fruit have an initial non-bearing period of a year for strawberries and more for cane fruit. With the increased investment necessary for perennial crops the economic risk is increased.

Intra-European trade tends to be higher than that for vegetables. This is occasioned by the distinct and natural climatic regions required by the production of different types of fruit. Also fruit is generally of higher value compared with vegetables where as this only applies to a much smaller proportion of vegetables produce. Salad vegetables, in particular glasshouse vegetables are on their own in economic character among vegetable production. Here, investment in glasshouses and heating is extremely high and the business is also highly labour intensive. While vegetables and fruit may have some different economic characteristics, both kinds of produce are required to be grown to perfection as the EEC quality standards are high. This leaves the producer with the responsibility of achieving quality and quantity to survive in the market for his produce. While an element of competition is from a distant supplier, the competition from the produce next door is just as real, if not more so.

CHAPTER 3

TRADE POLICY FOR FRUIT AND VEGETABLES

The Philosophy Behind Trade Policy

Introduction

It is not surprising that the most important contribution on trade policy for fruit and vegetables, both in theory and in practice, should come from Holland, where respect for the land and the capacity to exploit it is not surpassed by any other country. In the Netherlands, with the most advanced horticultural sector we know, fruit and vegetables have been of particular importance as the means of livelihood of a significant proportion of the Dutch population for generations. The horticultural sector, apart from fruit and vegetables, includes ornamentals. All three types of production involve the intensive use of capital, land and labour, and unlike many agricultural products, they are sold in a free market. This sector was seen by the Dutch government in the depression of the 1930's as one of such national importance as to be given specific policy measures to support it.

Support came mainly in the form of cheap credit, which government authorised banks to administer, and in the development of the Dutch Auction system with powers to grade produce into quality classes, to reject produce when necessary and carry out market intervention in times of oversupply. The high quality standards of the Dutch produce became the pattern for quality standards in the development of EEC fruit and vegetable policy later on. Developments in maintaining high standards of production and quality of produce were assisted by scientists from the Dutch Ministry of Agriculture. The inevitable concentration of horticulture in the Netherlands gave the social cohesion necessary for the successful implementation of policy measures. Constant up-dating took place in relation to growing techniques by the majority of horticultural producers.

The Director of the Dutch Central Bureau of Auctions gave a paper in 1975 on horticultural policy, at the Symposium of the economics section of the International Association of Horticultural Science, which encapsulated contemporary philosophy on horticultural policy in Western Europe[1]. The paper recognised that price is the main regulator of economic activity, and that the two main pivots are demand and supply, and through price, income.

The Function of Price and Objectives of Horticultural Policy

On the function of price in the objectives of horticultural policy, Kemmers saw that since the fifties the income of producers has been the main objective of governments and produce organisations throughout the capitalist world. A second important objective, particularly of government policy, was always the increase in production for the benefit of general welfare. Perhaps the safeguarding of

---

[1] Kemmers W.H. Prices and Economic Policy in Western Europe. Acta Horticulturae No 55 pp 33-9, 1976.

consumer requirements has occasionally been an objective of policy too. In Kemmers' words, however, safeguarding the consumer related to the supply of fruit and vegetables, and in no sense to price. To what extent does price formulation contribute to these objectives? The answer to this question cannot be assessed in reality because in no country is the price formation process completely free. Reality is the result of the regulatory effect both of price and policy. The provision of consumers in Western Europe is at a high level and the contribution of horticulture to the general welfare has grown in step with economic expansion in many countries. Horticultural incomes fluctuate more than those in other sectors of the economy. As a result of economic pressure many horticultural producers have been squeezed out. At the same time many producers have been able to achieve satisfactory incomes because of successful adjustments in their business management. The range in profitability between individual producers of fruit and vegetables in a given financial year is well known. This has been demonstrated over the years by the Cambridge University annual reports on economic results from horticulture.[2] The achievement of a satisfactory income, which, detachedly, is good as it reflects competitive entrepreneurial activity, is not a result of policy only. The general economic dynamism of which price formation in horticulture is a part also has also been a definite influence.

Long run factors which raise horticultural incomes and increase supply can bring dangers of considerable price falls and reduction in incomes. Simultaneous demand expansion offsets possibilities of price falls and income reduction. Difficulties with regard to income levels arise where demand fails to grow sufficiently, so the increase in producers' income is less than that in other economic sectors. For producers to leave horticulture is acceptable under the dynamic economic process of the industry. There is not room for every horticultural business to enjoy increasing prosperity but there is room for expansion in those horticultural businesses with potential.

Price Fixing and Income Support

An incomes policy can be based on a minimum or intervention price. Produce not sold at this price can be taken off the market. Such a measure prevents further price falls in a period of abundant supply. If demand is inelastic this measure prevents a further decrease of returns and income. Any compensation from the public funds for produce withdrawn from the market can be added to income. Supplementing income from public funds can be applied on a producer basis. Methods are an addition to the price or a supplement per acre. An income supplement can also be limited to a certain type of enterprise. Possibilities are social amenities which, within a certain sector of society, particularly the agricultural society, are supplied free of costs or more cheaply than in other sectors. An alternative is the application of fiscal systems, for instance the system of taxation according to acreage, type of crop, or type of agricultural occupation. These leave incomes largely untaxed, as opposed to incomes in other sectors. High taxes sometimes stimulate investment; expenditure, for instance on equipment, may be offset against tax, though an economic climate favouring investment must be there. Producers expand their business in the good years up to a point, instead of paying taxes. This can have a positive effect in the long run in that businesses keep up to date and so preserve their earning capacity.

Measures Affecting Supply

An incomes policy to limit supply, thus raising price and income, may be carried out in various ways such as import restrictions and export subsidies or processing produce, which usually amounts to giving a subsidy to the producer concerned or restricting production. A policy to increase production or to prevent a reduction of production for the benefit of general welfare will be able to use the latter methods, expecting that a higher income will cause an expansion of production or will prevent a reduction. Other methods, if used as well, such as reducing costs and raising income by subsidising costs and cheap credit, may be practised for a while, with an ultimate alternative of transferring horticultural producers to other profitable industries. Direct subsidies on investment are associated with

---

[2] Economic Results from Horticulture. Hinton, Lynn and Housden, Walter. Department of Land Economy. Cambridge University Series.

this policy. Finally, as part of the policy, there is provision for scientific research, education and an advisory service. The latter has become part of policy in most advanced countries, and it works well as a benefit to those fruit and vegetable producers who have the management ability to take advantage of it.

The Influence of Policy Measures to Stimulate Incomes

The producers' income is mostly the target of policy, not the price level itself. Only for safeguarding market supply is the price level sometimes the objective. If the policy objectives do not concern the income of producers but aim at an expansion of the volume of production, policy is based on the supposition that the resulting downward price tendencies will not affect income. There are some arguments in this connection. It may be that the influence of stimulated home supplies on the general price level is negligible and a growing demand to absorb the increasing supply without price falling is expected. It may be that home production is sufficiently competitive and that elsewhere production will disappear, thus preventing a fall in price and income. An increased income can induce a large supply which presses price downwards and may provoke a political urge to use more powerful instruments. Policy measures can raise income again to be followed by action on the supply side as before.

While the political decision making is a slow process, the price mechanism has often caused income effects before policy measures can undo them. This means that the effect on supply is weakened and producers continue to be aware that their business entails price and income risks. At this point policy measures stimulating production can be risky. Measures of policy are generally designed when the price level promises a good income. Then the regulatory effect of the price mechanism induces increases in supply. If the influence of implementation of production stimulating policy measures comes on top of that in a subsequent stage, over-supply and a considerable price falls are inevitable. This is reduced by the fact that the flow of people from horticulture, even when it is under pressure, is smaller than might be expected. A high income may have the effect of slackening of efforts. Technical progress means that expansion in horticulture is often greater in those areas which experience low incomes, where modernisation promises an improvement.

The Needs of Horticultural Policy

The overall contention is that price alone as a regulator of economic activity will not bring harmony between production and demand. A horticultural policy is required to protect producers from extreme risks. No horticultural policy will attain harmony and thus stability. Both horticulture and the general economic framework in which horticulture functions, are subject to too many disturbing factors. If these disturbances originate from short-term fluctuations in supply and demand, the need for horticultural policy measures will continue to be limited. In particular long run changes need policy measures. It is possible to remove extreme risks through well directed policy measures without disregarding trends.

Kemmers' philosophy of trade policy for fruit and vegetables would serve to protect the income of producers to a certain degree. However, the trade policy of the EEC in the fruit and vegetable sector, which was developed between 1962 and 1972, gave more protection to the sector than Kemmer himself had in mind. The Dutch experience in trade policy for fruit and vegetables greatly influenced the trade policy developed by the EEC.

Trade Policy for Fruit and Vegetables

International trade in fruit and vegetables depends on a viable demand in advanced countries, where personal incomes are high enough to pay for costs of production to the "farm gate", and for the cost of packing and shipment to leave a profit margin at each stage. This does not suggest high profits in absolute, merely profits high enough to reward participants over and above the commercial risk of the business concerned. Fruit and vegetables are seldom, if at all, sold firm at the point of origin. The nearest the trade gets to firm selling is when the wholesalers concerned are panel members at the point of destination, appointed by exporting trade organisations. Such organisations, for example, are the South African Deciduous Fruit Board, the New Zealand Apple and Pear Marketing Board, and Agrexo which handles produce from Israel.

While the international trade in fruit and vegetables is largely determined by demand and supply, countries which represent big markets are precisely those which have most highly developed trade policies, designed to protect the incomes of their domestic producers. Such countries or country groups are Japan, the United States, the European Economic Community and the Comecon. Britain had her own highly sophisticated trade policy of Commonwealth preference before she joined the Community in 1973. By doing so of course Britain accepted the even more sophisticated and yet more protective policy of the European Economic Community. Because of the motive of Western governments and of country groupings to give preferential policy to their domestic producers of fruit and vegetables, as we near the year 2000 so these restrictive policies, restrictive to third and mostly poor countries, will now become an area for genuine political and economic debate. While the present situation is deplorable, it may become appreciated that a radical change in policy can increase the influence of the West in the Third World, a policy of vital importance to the West. It is uncommon for altruism and necessity to coincide.

Trade policy of the EEC is of considerable importance to international trade in fruit and vegetables because the Economic Community is the greatest world market[3]. This policy is to the great disadvantage of a number of countries overseas, all seeking the opportunity to develop economically, many of whom have potential in the export of fruit and vegetables. A telling example is the Common Customs Tariff of twenty per cent on bananas, not even grown significantly in the Community. The EEC trade policy for fruit and vegetables has been exclusively established to protect the EEC producers[4]. At the same time there are concessions on the EEC fruit and vegetables tariffs for certain Mediterranean countries, and for members of the African, Caribbean and Pacific trading group, the Lomé countries.

The United States and Japan have fruit and vegetable trade policies similar to that of the EEC and they are somewhat similar in their effects. However, the United States trade policy is more liberal to developing countries. All three policies are based on agricultural protection and together these economies represent almost the entire world market for fruit and vegetables. The Comecon trade policy is designed to supply the Soviet Union, chiefly from countries of Eastern Europe and Cuba. Events in Europe that occurred in late 1989, mean that one day, much sooner than expected, protection will be history, like the English Corn Laws of the last century. The increased free market of East Europe and the Soviet Union and the abandonment of protectionism may be expected to develop strongly over the next generation. When, as is likely, purchasing power in the Soviet Union increases, it will have the effect of shifting the balance of existing trade policies. Links with Eastern Europe from the market economy countries of Central and Western Europe resulting in the spread to the East of the concept of market economy, will sooner or later call for equality in trading between East and West Europe and some changes in the EEC trade policy for fruit and vegetables.

Professor Walter Holstein, the 'father' of the Common Market said some fifteen years ago "We are not in business but we are in politics". The basis for present trade policies for fruit and vegetables in the EEC, the United States and Japan is purely political. When a less selfish politics operates so trade policies for fruit and vegetables will be changed. It may be a phenomenon, but agricultural policy based on protection in the advanced countries suggests that domestic farmers should be given protection as of right, when other sectors of the economy are not. This must change if the developing world is to be wooed rather than be made to suffer because of the trade barriers established by the advanced economies.

A.H. Boerma, one time Director General of FAO, spoke of policies created by FAO as being founded on the conviction that their implementation would help to achieve greater equity in world trade and accelerate balanced economic development. Equity is missing in the trade sector of fruit and vegetables. These commodities are of course not major agricultural commodities, but because the policy for fruit and vegetables shadows agricultural protection in advanced countries, trade policy

---

[3] International Trade Policy for Fruit and Vegetables. Hinton Lynn Acta Horticulturae 97, pp 347-362 1979.

[4] "National Policies and Agricultural Trade: a study of the EEC" OECD Paris 1987

for these commodities moves against equity in world trade and also does little to accelerate balanced economic development.

EEC Trade Policy for Fruit and Vegetables

EEC trade policy for fruit and vegetables, started in 1962, was developed under Regulation 1035 of 1972. The high quality standards which had been practised in Holland over the years, were applied to a number of fruit and vegetable commodities in all member states. West Germany, France, Holland, Belgium and Luxembourg, and Italy are founder members (1957), Denmark, Britain and Ireland acceded in 1973, Greece acceded in 1980, and Spain and Portugal in 1986. Before 1973, West Germany and Britain together accounted for over half the imports of fruit and vegetables in Europe as a whole. At this time, imports of fruit and vegetables to the Community were a quarter of all supplies on the EEC market. Given such a large export market, the EEC trade policy for fruit and vegetables is of vital concern to exporting third countries in most parts of the world. For so many, better access to the EEC market would give them more of an opportunity for economic development. The Common Organisation of the Market in fruit and vegetables has been developed in the context of the Common Agricultural Policy which is based on agricultural protection.

Trade Concessions to Mediterranean Countries and to Developing Countries

The first agreement, as part of the global Mediterranean policy of the EEC, was with Israel in 1975. While it eased agricultural trade with Israel, only small reductions were made on tariffs on Israeli exports of fruit and vegetables, which were regarded by the EEC as competing products. In 1976 the Magreb countries, Morocco, Africa and Tunisia, were treated generously over agricultural products, but as with Israel, competing products such as fresh vegetables, grapes and citrus fruits were not treated so well. They attracted a tariff reduction of forty to eighty per cent. The agreement with Israel took account of EEC exports of industrial machinery. Agreements with Cyprus, the Mushraq countries, Egypt, Jordan, Lebanon and Syria in 1977, also gave less generous concessions on fruit and vegetables, with quantitative limits for example on citrus fruit. Agreements with Cyprus in 1978 were less generous then those in Magreb and Mushraq countries. For instance, though citrus fruits are of importance in Cyprus, the CCT reduction on oranges was only sixty per cent compared with eighty per cent for the other agreements.

During the first part of the 1970's the EEC, Japan and the USA made bilateral agreements, originally intended by UNCTAD as reverse preferences with a number of developing countries. The policies of UNCTAD overall, so far, has resulted in small reductions, if any, in the tariffs on fruit and vegetables in developing countries. At the end of 1990 agreement on the Uruguay Round was not achieved for the major agricultural commodities which the United States sought in vain. A more liberal global trade policy for fruit and vegetables is not likely to come about until trade in the major agricultural commodities is liberalised.

The 1986 enlargement of the European Community to include Spain and Portugal has serious repercussions for international trade in fruit and vegetables. For example, until 1986 Spain exported over a sixth of the fruit and vegetable imports to Britain, against the barrier of the CTC. Now the EEC markets will become freely open to Spain, a very important supplier, with a large range of produce. Spain will, of course, from now on divert more of her exports to the EEC and less will be available for third countries. The transition period for trade in fruit and vegetables means, in practice, that complete free trade for Spain and Portugal with other EEC member states is being phased over a ten year period.

The enlargement of the EEC has added considerably to its fruit and vegetable supply. Spain of course has great potential, Greece rather less and Portugal much less. If there is no relaxation of the EEC trade policy for fruit and vegetables, a number of third countries will be in genuine difficulties. Especially will this be so for East European exporters who need hard currency but the new independence of Eastern Europe has transformed the situation. The new freedom for Eastern Europe will bring EEC concessions, if not eventually free trade, with Eastern Europe in the fruit and vegetable sector. Towards the end of 1990, full economic and political union came about in the former East and West Germany which means one fruit and vegetable trade policy for Germany, that of the EEC.

Following the accession of Spain and Portugal in 1986, the EEC concluded Additional Protocols to the existing agreements with Mediterranean third countries, in order to prevent erosion of their traditional trade with the EEC.[5] Through the provisions of the Additional Protocols, the concessions to the Mediterranean third countries mirror the tariff position for Spain and Portugal, as they reduce in the ten year transition period. These concessions, however, are limited to tariff quotas, concessionary amounts based on a three year average of imports into the EEC, for certain commodities from each country. Tariffs reduce at the same rate and over the same period as for Spain and Portugal. In addition, certain commodities are bound by reference quantities which, although slightly more flexible, have the same practical effect as tariff quotas.

The Trade in Fruit and Vegetables Between Developed and Developing Countries

Exports of fresh fruit from developed countries are some fifty eight per cent of world exports, but those of processed fruit are much greater, seventy-five per cent. Fresh vegetable exports in developed countries are yet higher, seventy-one per cent, which along with fruit means that the fresh trade substantially belongs to developed countries. The low proportion of processed fruit in developing countries exports is primarily a function of the effect of the high tariffs imposed by developed countries on processed fruit and vegetable imports.

While bananas are the reserve of tropical countries, citrus fruit production is divided almost equally between industrial countries, the United States and Mediterranean countries, and developing countries. In exports the developing countries dominate marginally. Enlargement of the EEC to include the citrus fruit of Greece, Spain and Portugal, will have aggravated the trade inhibiting effects of Community policies on third country citrus producers.

World Energy Resources

An aspect of overwhelming importance to the world economy is the role developing countries can play, mainly tropical and sub-tropical countries, in producing fruit and vegetables in a natural climate. Temperate zone advanced countries absorb a considerable amount of scarce fossil fuel. Adjustment, however difficult, affecting producers in the production of glasshouse produce, and better use of warmer climates, would be to the advantage of economic development in warmer climates and of global fuel conservation. Trade policies need adjustment if this point is considered to be important by the governments concerned. Weighing fuel use in temperate climate heated crop production against fuel use in transport from countries with warm climates, should be given carried out.

The Costs of the Fruit and Vegetable Règime

The fruit and vegetable règime has been a relatively low cost règime, with current FEOGA spending at only four or five per cent of the agricultural budget. It has been even less for fresh fruit and vegetables.[6] Fruit and vegetables have not experienced the spending crises and political limelight of the sugar, dairy and cereals sectors. However, this situation is changing. In recent years the levels of intervention for some fruit and vegetables have been high. For example, during 1983-4, 60 per cent of all EEC mandarins were withdrawn from the market. For lemons the figure was 28 per cent, 17 per cent for oranges, 14 per cent for peaches and 13 per cent of apricots. Also in recent years more apples have been taken off the market. Such surpluses have led to the introduction of a stabiliser règime for a number of commodities. On the other hand, enlargement, which will increase both the political and economic significance of this sector, may also put greater pressure on market support spending for fruit and vegetables, both in terms of greater volume involved, and by the likelihood of calls for increased protection by the strengthened Mediterranean representation in the EEC Council. The latter is less likely now that a stabiliser policy has been introduced, a definite reversal of the policy of protection.

---

[5] The countries concerned are Algeria, Cyprus, Egypt, Israel, Jordan, Malta, Morocco, Tunisia, Turkey, Yugoslavia, the Lebanon and the Canary Islands.

[6] The EEC market policy for fruit and vegetables. Avermuete U Acta Horticulturae No 223 pp 37-40 1987.

The Common Agricultural Trade Policy for Fruit and Vegetables

Introduction

As mentioned, the fresh fruit and vegetable règime was consolidated in 1972 under Regulation 1035/72. The Common Organisation of the Market has the objective of achieving balance between supply and demand at prices which are fair to producers, encouraging specialisation within the Community and taking account of trade with third countries. It applies to all fruit and vegetables grown in the Community with the exception of potatoes, wine grapes, beans and peas for fodder and olive oil. Not included are bananas for which no common market organisation exists. The fresh fruit and vegetable sector is characterised by distinct fluctuations in price and in supply and demand for products which are highly perishable. The CAP règime for the sector consists of quality standards, internal market support and frontier protection.

The Internal EEC Règime of Market Support for Fruit and Vegetables

Quality Standards

Quality standards for fresh horticultural produce apply to some 30 different fresh fruit and vegetables. They define grading and labelling requirements and are applicable at all stages of distribution from the grower's packhouse to the retail shop and to imports and exports, but excluding farmgate sales and produce destined for processing. The aim is to keep products of unsatisfactory quality off the market, to guide production to meet consumers' requirements and to facilitate trade based on fair competition and thereby help to improve profitability.

The Market Intervention or Withdrawal System of EEC Market Support for Fruit and Vegetables

The objective of market intervention in this sector is to provide a safety-net in years of glut rather than create a guaranteed market. It takes the form of compensating producers from Community funds for withdrawing produce of marketable quality from the market, after which it can be distributed free to charitable organisations, hospitals, old people's homes, schools, prisons and so on, where this does not adversely affect the market. Alternatively, it can be disposed of in the form of animal feed or distilled into industrial alcohol. In practice, a substantial proportion of produce is wasted, i.e. dumped or ploughed back into the land. Intervention, or withdrawal prices are derived from the basic and buying-in prices set by the Council of Ministers as part of the annual price-fixing for the products concerned. Under Regulation 1035/72 on the Common Organisation of the Market in fruit and vegetables, the Council is required to take into account the need to contribute to the support of producers' incomes, to stabilise markets without leading to structural surplus, and to consider the interests of consumers. Buying-in prices must be set within fixed percentages of basic prices, 30 to 40 per cent in the case of cauliflowers, tomatoes and aubergines, 40 to 55 per cent for the other products concerned. The basic and buying-in prices are shown for selected commodities which qualify, and for the relevant marketing periods in Table 1. Although there is provision for Government agencies to organise withdrawal at the level of the buying-in price for certain products under certain conditions, withdrawal operations are generally carried out by recognised producer groups. As an incentive to cooperation, the buying-in price is adjusted by 10 per cent of the basic price. To calculate the withdrawal price actually paid to producers the buying-in price is adjusted by conversion factors determined through Management Committee procedure which are applied as appropriate in respect of different qualities, presentations, sizes and varieties to reflect relative market values.

In order to make processing a more attractive alternative to withdrawal, aids are paid to processors of lemons and certain varieties of oranges who agree to pay a minimum price to producers. This aid is fixed by the Management Committee and an element of the minimum price, which is usually changed annually by the same percentage as the withdrawal price. In order to encourage the consumption of Community produced fruit there are also marketing premiums on exports of Community produced class I oranges and mandarins to other member states at levels determined by the Council, and grants to encourage conversion to citrus varieties more suited to consumer demand. Export refunds are available to Community exporters of certain produce to third countries.

Table 1. Basic and Buying-in Prices for Selected Fruit and Vegetables - ECU per 100 kg

Commodity	Basic price		Buying-in price	
	EEC10	Spain	EEC 10	Spain
Mandarins				
16 Nov 90 to 28 Feb 91				
16 to 30 Nov	37.01	28.46	23.68	18.18
Descending to February	34.47	26.84	21.90	17.04
Sweet oranges				
1 Dec 90 to 31 May 91				
December	34.51	31.47	21.82	19.89
Falling and rising to April and May	34.12	31.15	21.06	19.28
Grapes				
1 Aug to 20 Nov 90				
August	36.31	36.31	23.35	23.35
Sep to Nov	32.51	32.51	19.92	19.92
Pears				
1 Jul 90 to 30 Apr 91				
July	28.67	20.01	14.75	10.31
Falling and rising to Jan to Apr	27.65	19.35	14.62	10.23
Peaches				
1 Jun to 30 Sep 90				
June	45.38	43.18	25.21	23.99
Jul to Sep	42.99	41.03	24.08	22.98
Cauliflowers				
14 May 90 to 30 Apr 91				
May	30.96	24.43	13.47	10.62
Falling and rising to April	29.67	23.46	12.92	10.21
Tomatoes				
11 Jun to 30 Nov 90				
11 to 20 Jun	25.45	15.46	10.82	5.88
Falling and rising to November	28.32	15.40	11.34	6.12
Aubergines				
1 Jul to 31 Oct 90				
July to October	17.77	9.80	7.12	3.65

There is also basic and buying in for apples, satsumas, clementines, nectarines, apricots and lemons.

Source: Brussels 1990

Table 2. The Common Customs Tariff - Selected Vegetables and Principal Concession Beneficiaries

Commodity	Per cent	Concessions to:-
Potatoes		
1 Jan to 15 May	15	ACPOCT, LDDC, TR free; Sp 5.6%
16 May to 31 Dec	21	ACPOCT, LDDC, TR free; Sp 7.8%
Tomatoes		
1 Nov to 31 Dec	11 Min 2 ECU/100 kg	MGB,OT 4.4% Min 0.8 ECU/100 kg, Sp 3.8% Min 0.3 ECU/100 kg, TR free
1 Jan to end Feb	11 Min 2 ECU/100 kg	MGB,OT 4.4% Min 0.8 ECU/100 kg, Sp 1.9% Min 0.3 ECU/100 kg, TR free
1 Mar to 31 May	11 Min 2 ECU/100 kg	MGB,OT 4.4% Min 0.8 ECU/100 kg, Sp 3.8% Min 0.7 ECU/100 kg, TR free
1 Apr to 14 May	11 Min 2 ECU/100 kg	MGB,OT 4.4% Min 0.8 ECU/100 kg, Sp 6.3% Min 0.7 ECU/100 kg, TR free
Onions		
15 Feb to 15 May	12	ACPOCT, MGB, OT 4.8%; TR free
1 Feb to 10 Apr	12	LDDC 6.0%; TR free
Cauliflowers		
15 Apr to 30 Nov	17 Min 2 ECU/100 kg	PT 6.6% Min 0.2 ECU/100 kg; TR free
1 Dec to 14 Apr	12 Min 2 ECU/100 kg	PT 6.6% Min 0.2 ECU/100 kg; TR free
Lettuce		
1 Apr to 30 Nov	nil	
1 Jul to 30 Sep	15	ACPOCT 6.6% Min 1.1 ECU/100 kg; Sp 5.2% Min 0.8 ECU/100 kg; TR free
Carrots		
1 Jan to end Feb	17	
1 Mar to 31 Mar	17	ACPOCT 9.2%; CY 4.9%; TR free
1 Apr to 31 Dec	17	Sp 9.2%; TR free
Cucumbers		
1 Nov to 31 Dec	16	ACPOCT 7.1%; Sp 5.6%; TR free
1 Jan to end Feb	16	Sp 5.6%; EG 7.1%; TR free
1 Mar to 15 May	16	ACPOCT 7.1%; Sp 5.6%; TR free
Asparagus		
15 Aug to 30 Sep	16	ACPOCT, Sp 8.7%; PT 7.1%; TR free
1 Oct to 31 Oct	16	ACPOCT, Sp 8.7%; GSP 12%; LDDC free TR free
1 Nov to 31 Dec	16	ACPOCT, Sp 8.7%; GSP 12%; LDDC free TR free
Mushrooms		
all year	18	ACPOCT free; Sp 6.7%; LDDC, TR free; PT 5.1%

Source: Brussels 1990

Table 3. The Common Customs Tariff - Selected Fruit and Principal Concession Beneficiaries

Commodity	Per cent	Concessions to:-
Strawberries		
1 May to 31 Jul	16 Min 3 ECU/100 kg	SP 8.7% Min 1.6 ECU/100 kg; PT 7.1% Min 1.3 ECU/100 kg; TR free
1 Aug to 30 Sep	14	SP 7.6%; PT 6.2%
1 Oct to 30 Oct	14	SP 7.6%; PT 5.2%
1 Nov to end Feb	14	CY 4%; SP 6%; MGB 5.6%; PT 5.2%; TR free
1 Mar to 31 Mar	14	CY 4%; SP 7.6%; MGB, MT 5.6%; TR free
1 Apr to 30 Apr	14	SP 7.6%; PT 6.2%
Raspberries		
15 May to 15 Jun	11	SP 5.9%; GSP 9%; LDDC,TR free;MGB 5.5%
16 Jun to 14 May	11	SP 5.9%; GSP 9%; LDDC,TR free;PT 6.6%
Papaws		
	2	ACPOCT,LDDC,GSP, TR free; PT 1.3%
Apples		
1 Aug to 31 Dec	14 Min 2.4 ECU/100 kg	SP 4.9% Min 0.8 ECU/100 kg
1 Jan to end Feb	8 Min 0.8 ECU/100 kg	SP 2.8% Min 0.8 ECU/100 kg; PT 3.5% Min 1 ECU/100 kg; TR free
1 Mar to 31 Mar	8 Min 2.3 ECU/100 kg	SP 2.8% Min 0.8 ECU/100 kg; PT 3.5% Min 1 ECU/100 kg; TR free
1 Apr to 31 Jul	6 Min 1.4 ECU/100 kg	SP 2.1% Min 0.4 ECU/100 kg; PT 2.6% Min 0.6 ECU/100 kg; TR free
Oranges		
1 Apr to 30 Apr	13	ACPOCT 2.6%; CY 3.7%; EG 2.7%; MGB 2.6%; TR free
1 May to 15 May	6	ACPOCT 1.2%; CY 1.7%; MGB 1.2%;TR free
16 May to 15 Oct	4	ACPOCT 0.8%; CY 1.1%; SP 0.8%; EG 1.6%; MGB 0.8%; TR free
16 Oct to 31 Jan	20	ACPOCT 4%; CY 5.8%; EG 8%; IL,JO,LB 8%; TR free
1 Feb to 31 Mar	20	ACPOCT 4%; CY 5.8%; EG 8%; IL,JO,LB 8%; TR free
Lemons		
15 Jan to 14 Jun	8	CY,EG 3.5%; IL,JO,LB,MGB 4.8%; TR free
15 Jun to 14 Jan	8	CY,EG 3.5%; SP,MGB 1.6%; PT 3.5%; TR free
Bananas		
all seasons	20	ACPOCT free; SP 7.5%; PT 5.7%; TR free
Pineapples		
All seasons	9	ACPOCT free; SP 3.3%; PT 0.7%; LDDC,TR free

Table 3 - continued

Commodity	Per cent	Concessions to:-
Peaches		
1 Dec to 31 Oct	22	SP 7.7%; PT 9.7%; TR free
1 Jan to end Feb	22	SP 7.7%; PT 9.7%
1 Mar to 31 Mar	22	ACPOCT 9.7%; SP 7.7%; PT 9.7%
1 Apr to 30 Nov	22	SP 7.7%; PT 9.7%
Grapes		
1 Nov to 14 Nov	18	SP 6.3%; PT 8%
15 Nov to 30 Nov	18	SP 6.3%; MGB 7.2%; PT 8%; TR free
1 Dec to 31 Dec	18	EG,MGB 7.2%; SP 6.3%; PT 8%; TR free
1 Jan to 31 Jan	18	EG, LB 7.2%; SP 3.1%; MA,PT 4%; TR free
1 Feb to 31 Mar	18	DZ,EG,MA,PT 4%; SP 3.1%; TR free
1 Apr to 30 Apr	18	EG 7.2%; SP 6.3%; JO,MA,PT 4%
1 May to 7 Jun	18	EG,JO 8%; SP 6.3%; MGB 7.2%
8 Jun to 17 Jun	18	EG,IL,JO,PT 8%; SP 6.3%
18 Jun to 14 Jul	18	EG,IL,JO,PT 8%; SP 6.3%; TR free
1 Jul to 14 Jul	18	PT 8%; SP 6.3%; TR free
15 Jul to 17 Jul	22	SP 7.7%; PT 9.7%; TR free
18 Jul to 31 Jul	22	SP 7.7%; PT 9.7%; TR free
1 Aug to 4 Aug	22	SP 7.7%; PT 9.7%; TR free
5 Aug to 31 Oct	22	SP 6.7%; PT 8%; TR free
1 Nov to 31 Dec	18	SP 6.7%; PT 8%; TR free

Source: Brussels 1990

The External EEC Règime of Market Support for Fruit and Vegetables

Tariffs

The Common Customs Tariff (CCT) applies to the full range of fruit and vegetables, whether they are subject to intervention and/or reference prices (see below) or not. Levels are generally higher in periods of Community production and lower in winter. There are some concessionary rates for third countries through the Lomé, Generalised System of Preference and Mediterranean third country agreements, and also for the Canary Islands, but these are normally limited by tariff quota and often by calendar period. The CCT and concessionary rates of EEC duty on selected fruit and vegetables are given in Tables 2 and 3. Tariffs are seen to range from two to twenty per cent, and ACPOCT and LDDC are free of tariff or have a reduced tariff for several commodities and for several periods of supply. Turkey is generally free of tariff.[7]

The Reference Price or Minimum Import Price System

The reference price operates to protect Community producers during their main marketing periods from low priced imports from third countries. It applies to the sensitive products listed in Table 4 and is distinct from, and additional to, the operation of the CCT. It is effectively a minimum

[7]Turkey, Horticulture and the European Community, Bath University 1988, Sergeant, Malcolm.

Table 4. The EEC Reference or Minimum Import Prices, Ecu per 100 kg

	Ecu per 100 kg		Ecu per 100 kg
Apples 1989/90		**Pears 1990**	
July	46,25	July	46,78
August	40.86	August	39,72
September	43,98	September	39,42
October	43,78	October	42,59
November	44,61	November	45,65
December	45,66	December	48,77
January	48,62	January to April	51,01
February	50,44		
March	52,73		
April	54,00		
May	56,31		
June	57,08		
Cucumbers 1990		**Plums 1990**[a]	
11 to 20 February	144,61	11 June to 31 July	69,39
21 to 28 February	122,42	August	69,39
March	112,14	September	60,71
April	92,76		
May	76,12		
June	63,76		
July	48,36		
August	48,73		
September to 9 October	57,72		
10 October 10 November	81,6		
Lemons 1990/91		**Peaches 1990 including nectarines**	
June	54,59	11 to 20 June	82,58
July and August	60,82	21 to 30 June	71,65
September	56,33	July	71,22
October	50,57	August	56,98
November to April	47,15	September	56,23
May	47,73		
Cherries 1990			
21 to 31 May	140,95		
June	125,92		
July	115,69		
1 to 10 August	88,73		
Aubergines		**Courgettes (from Spain) 1990**	
April	87,72	21 to 30 April	
May	82,00	May	63,12
June	78,58	June	39,02
July	69,85	July	38,63
August	46,72	August	44,99
September	50,09	September	49,68
October	54,13		

[a] certain varieties have a lower reference price

Table 4 - continued

	Ecu per 100 kg		Ecu per 100 kg
Courgettes (from other Countries) 1990		**Artichokes** 1989/90	
21 to 30 April	71,79	1 Nov to 31 Dec	89,38
May	63,12	1 Jan to 30 April	78,83
June	42,18	May	74,95
July	38,63	June	63,53
August	44,99		
September	49,68		
Endives 1989/90			
from 15 Nov to 31 Jan	58,79		
1 Feb to 31 March	63,44		
Apricots (from Spain) 1990		**Apricots (other countries)**	
June 1 to 10	99,46	June 1 to 10	106,26
11 to 20	82,26	11 to 20	93,94
21 to 30	68,75	21 to 30	82,07
July	69,42	July	73,15
Lettuce 1989/90		**Sweet Oranges** 1989/90	
Nov to 31 Dec	70,35	1 Dec to 11 May	22,66
1 Jan to 28 Feb	75,601		
1 March to 31 May	82,34		
Tomatoes 1990/91		**Clementines** 1989/90	
April	197,27	1 Dec to 28 Feb	59,57
May	136,75		
1 June to 10 July	99,96		
11 July to 31 Aug	91,97		
September	45,07		
1 October to 20 December	46,55		

Source: Brussels 1990

import price system whereby, when prices of a particular product from a particular country representing a significant proportion of imports on a representative Community market fall below the reference price plus the full rate of CCT for two days, countervailing charges are applied to make up the difference. The countervailing charges refer only to a particular country and have a cumulative effect as long as produce continues to come in below reference price levels. The charges are not removed until prices are above reference price levels for two consecutive market days or when no prices are recorded for six consecutive market days. The system differs from most other CAP règimes in that the effect is to discourage imports below the minimum import price rather than to bring the price of imports up to institutional levels by the addition of levies. Reference prices are determined by the Management Committee and the present method is to take the lower of the two following calculations. Firstly, average producer prices in the three preceding years plus the increase in production costs in the latest year, less an allowance for increased productivity. Secondly, the reference price in operation in the previous year plus the increase in production costs, less an allowance for increased productivity. In practice the latter is usually the lower of the two and the result is that reference prices are generally below current producer prices.

The reference price is superficially like any other minimum import price of the CAP. In all member states prices of third country imports are monitored by the Commission. Even for those countries which have imposed preferential rates on the tariff (CCT) on their exports to the Community, the calculation to determine whether the reference price is being respected involves the full rate of the CCT. All exporting countries for any particular product face the same minimum import price, below which produce should not be able to enter the EEC. Under this system a tariff concession granted by the EEC does not necessarily lead to a price advantage with respect to foreign supplies in the EEC market. The reference price system might lead to non-price competition, more stringent grading, better packaging and promotion, all with the aim of raising quality, and the reference price system is more likely to act as a trade barrier to those third countries who compete on price rather than quality.

The Reference Price System for Citrus Fruits.

Citrus fruits are treated differently from other produce. The policy for citrus fruits which did not work particularly well when it was introduced in the 1960s led the Community to establish and assist the export of top quality Community citrus fruit from one member state to another. From 1975 onwards payments known as market penetration or marketing premiums were paid on such trade to help placement of citrus fruit. Transport costs were included in these premiums and so annual increases were made to the marketing premium rather than to the reference price. In 1982 legislation was passed by the Council to phase out the market penetration price system applicable to all producers. These changes are causing an upward drift in reference price levels for citrus products.

The Impact of the Reference Price System

In seeking the effect of the EEC reference prices on the market for fruit and vegetables, Williams and Ritson put commodities into four groups[8]. The first group, aubergines, Cox apples, Williams pears, Comice pears and Conference were found to be typically priced well in excess of the reference price and incurring no countervailing charges. Their market prices do not appear to be influenced by reference prices. The second group, plums, courgettes, cherries and table grapes, all tend to have a short production and reference price season with typically high prices which fall to levels close to the reference price briefly at the height of the EEC season. So there may be some potential for countervailing charges and a protective effect. The third group comprises cucumbers, tomatoes, Golden Delicious apples and peaches. Wholesale prices of these commodities do appear to influenced by the reference price and for which on occasions import prices dip below it and countervailing charges are incurred. The fourth group comprises citrus fruit, lemons, mandarins, oranges and clementines. Currently the reference price provides little or no impediment to third country imports but for which it is predicted that reference prices will become more active in future.

Overall the study showed that in the case of tomatoes, cucumbers, Golden Delicious apples and peaches, reference prices do appear to be influencing the UK market. The evidence is strongest in the case of tomatoes and cucumbers. There were indications that at the beginning of the reference price season, produce from Spain (before accession) and the Canaries is first offered at higher prices, with countervailing charges being incurred when market prices drop and Spanish produce gradually disappears from the market. The Spanish season would be ending at this time but its disappearance does appear to be hastened by countervailing charges rising to prohibitive levels. This applies to the situation before 1986 when Spain was not a member of the EEC. To a lesser extent the reference price is also active at the end of the season. It seems that UK consumers are being denied for a part of the year the opportunity to purchases of produce from the Canaries and Spain at competitive prices. Also that when Spanish transition is complete and Spain no longer has to respect reference

---

[8] The Impact of the EEC's Reference Price System on the Marketing of Fruit. Williams, Helen and Ritson, Christopher, University of Newcastle-upon-Tyne 1987.

prices, British, and more so Dutch, growers of tomatoes and cucumbers will face increasing competition from Spain.

Another important conclusion of the study by Williams and Ritson is that the change in the method of calculating reference prices for citrus fruit will soon begin to push up market prices in UK. Those countries which are unable to control the marketing of their produce will find it increasingly difficult to market to the UK and eventually we may expect to see Spanish produce taking a larger share of the UK market by undercutting the third country reference price. The implications of the change in calculating the reference prices are substantiated by a case study relating to Cyprus.[9] However, it is also pointed out that in contrast, because EEC produce does not have to respect reference prices and is in some instances supported by the withdrawal mechanism, EEC producers are encouraged to, pay attention to quality in marketing.

Fruit and vegetables represent an important sector of agricultural production in the EEC of 10 - which existed to January 1986 and their significance increased with accession of Spain and Portugal. Fruit and vegetables are 13.7 per cent of the agricultural output of the 12, with only milk (19.2 per cent) and beef and veal (14.1 per cent) being of greater importance. Spain's fruit and vegetable sector is relatively well organised and low cost, especially the citrus sector with many production and marketing cooperatives. Spain is a major exporter of citrus, tomatoes, strawberries and onions. The only sector of significance in Portugal is tomato production for tomato paste. Although the transition period agreed with the new member states is substantial (10 years after 1986 for fruit and vegetable products) there are widely held fears within other EEC and in non-EEC countries that their produce will be pushed out of the EEC markets once Spain no longer faces tariffs and other import controls.

The EEC Market Intervention System of Market Support in More Detail

The arrangements in the EEC for the withdrawal from the market of certain fresh fruits and vegetables follow. The Intervention Board or its equivalent in each member state is the official government appointed department responsible for administration of this system. The EEC regulations concerned permit recognised producers' organisations or producer groups to withdraw produce from the market and to claim compensation provided conditions of presentation and disposal are met. The arrangements apply to a very wide range of commodities.[10]

A basic condition of intervention payments is that the withdrawal arrangements apply only to those fruits and vegetables which are marketed through recognised producers' organisations. This puts countries like the UK, where only 25 per cent of the fruit and ten per cent of the vegetables are sold through producer groups, at a disadvantage. In Holland, for instance, 90 per cent of the fruit and vegetables are sold through producer organisations. A producers' organisation means any organisation of fruit and vegetable producers which is formed on the initiative of the producers themselves. Such an organisation promotes the concentration of supplies and the stabilisation of prices at the producer stage. Another responsibility of the producer organisation or producer group is that it makes available to its members adequate technical means for presenting and marketing their produce. A producer group ensures that members, subject to some minor exceptions, sell through the organisation the total output of the product or products for which they are members. Further it requires from its members to improve the quality of their products and to adapt the volume of supplies to meet market requirements and to provide information on harvests and supplies. When the main EEC policy for fruit and vegetables came on stream in 1972, producer groups were given financial support to establish them.

The Recent Stabiliser Règime and the Apple Surpluses

Intervention thresholds for the withdrawal system are fixed by the Council. If the threshold is exceeded, basic and buying in prices for the product in question will be reduced in the following year on a progressive scale in tranches of 1 per cent up to a maximum of 20 per cent. The Council deter-

---

9 Future Prospects of Cyprus Exports to the EEC. Wiliams, Helen, University of Newcastle upon Tyne, 1986.

10 Acknowledgement is made to The Intervention Board for Agricultural Produce.

mines the amount of the excess for each given product which triggers each 1 per cent price cut. The Commission establish whether the thresholds have been exceeded. Thus the Commission have been given a positive control with which to contain surpluses of fruit and vegetables and to take the opportunity of saving support funds.

A stabiliser scheme was first introduced in 1987 for tomatoes, to reduce expenditure on market intervention. Article 16b of the EEC Regulation 1035/72 makes provision for the introduction of "Intervention thresholds". Stabilisers are the price control mechanisms for the open market, whereas Intervention thresholds are applied to intervened or withdrawn produce. The Commission will lay down the threshold each year per product and if more than this quantity is withdrawn, a percentage reduction in price for the following year is triggered. Where the market in a listed product is affected or is likely to be affected by imbalances giving rise or liable to give rise to the buying in of significant quantities, an intervention threshold shall be fixed, before the beginning of the market year. For those products and if intervention exceeds the threshold by a quantity bought in during one marketing year or the average quantity bought in in several marketing years, financial liability shall be borne by the producers. The stabiliser règime, is a complete reversal of trade policy for fruit and vegetables from one of protection to one of less protection. It is similar to the co-responsibility levy on cereals in that the producer is penalised on his production. Twelve commodities have been given a stabiliser règime. For apples, mandarins, peaches, oranges and lemons the intervention threshold has been applied to a decreasing proportion of the average production over the last five years. As mentioned, the penalty is paid by the producer in the following year that the threshold price is exceeded.

For satsumas, clementines and nectarines their threshold price is related to ten per cent of the average production in the last five marketing years. For tomatoes there is a threshold limit of 390 thousand tonnes and for cauliflowers the threshold limit is three per cent of the marketed production in the last five marketing years.

The Fruit and Vegetable Trade Policy of the United States

Introduction

There is some general legislation relating to the fruit and vegetable trade policy of the USA. Imports of fruit and vegetables into the United States are subject to the provisions of the Agricultural Marketing Act of 1937, the Plant Quarantine Act, the Federal Insecticide, Fungicide and Rodenticide Acts, and the Food, Drug and Cosmetics Act. The US International Trade Commission assigns specific rates of duty on imported fruit and vegetables under the Tariff Act of 1930.

The Agricultural Marketing Agreement Act 1937

The above Act requires that whenever the Secretary of Agriculture issues, grade, size, quality or maturity regulations under domestic marketing orders for certain commodities, the same or comparable regulations on imports of those commodities must be issued. Import regulations apply only during those periods when the domestic marketing orders are in effect. Currently, the following imported commodities are subject to import regulations, avocados, dates (other than dates for processing), grapefruit, table grapes, limes, olives (other than Spanish style olives), onions, oranges, Irish potatoes and tomatoes. Import regulations are amended from time to time to conform with changes in domestic marketing order regulations. The Marketing Order Programme of the United States is given later. This programme is related to the fruit and vegetable trade policy of the United States given in the Harmonised Tariff Schedules for the United States.

The Plant Quarantine Act

The Plant quarantine regulations of the US Department of Agriculture are divided into two classes, prohibiting and restrictive. Prohibitory orders forbid entirely the entry of designated plants and plant products, which are subject to attack by plant pests, and for which there is no treatment available that would ensure complete freedom from such pests. Restrictive orders allow the entry of plants or plant products either under a treatment requirement or an inspection requirement. Import permits are required. These permits are issued to resident importers in the United States, who are responsible for carrying out the conditions of entry under the permit.

The Tariff Act of 1930

Rates of duty on imported commodities are published periodically by the US Trade Commission in the Harmonised Tariff Schedules of the United States. Duties for selected vegetables and fruit are given in Table 5 and Table 6.

Table 5. The Harmonised Tariff Schedule of the United States for Selected Vegetables, 1990

Commodity	Units	General Duty	Special Duty	Rate of Duty 2
Jicamas and pumpkins	kg	25%	Free A,E,IL; 20% CA	50%
Potatoes	kg	17.5%	Free E,IL; 14% CA	35%
Peas - entering Jul 1 to Sep 30	kg	2.2¢/kg	Free A,E,IL; 1.76¢/kg CA	8.6¢/kg
- other	kg	4.4¢/kg	Free A,E,IL; 3.5¢/kg CA	8.6¢/kg
Spinach	kg	17.5%	Free E,IL; 14% CA	35%
Onions	kg	3.9¢/kg	Free A,E,IL; 1.2¢/kg CA	5.5¢/kg
Tomatoes - entering Mar 1 to Jul 14 or from Sep 1 to Nov 14	kg	4.6¢/kg	Free E,IL; 3.5¢/kg CA	6.6¢/kg
- entering Jul 15 to Aug 31	kg	3.3¢/kg	Free E,IL; 2.5¢/kg CA	6.6¢/kg
- entering Nov 15 to end Feb	kg	3.3¢/kg	Free A,E,IL; 2.6¢/kg CA	6.6¢/kg
Brussels sprouts	kg	25%	Free A,E,IL; 10% CA	50 %
Carrots	kg	1.1¢/kg	Free A,E,IL; 0.8¢/kg CA	8.8¢/kg
Cucumbers - entering Dec 1 to end Feb	kg	4.9¢/kg	Free E,IL; 3.9¢/kg CA	6.6¢/kg
- entering Mar 1 to Apr 30	kg	6.6¢/kg	Free A,E,IL; 5.2¢/kg CA	6.6¢/kg
- entering May 1 to Jun 30 or Sep 1 to Nov 30	kg	6.6¢/kg	Free E,IL; 5.2¢/kg CA	6.6¢/kg
- entering Jul 1 to Aug 31	kg	3.3¢/kg	Free E,IL; 2.6¢/kg CA	6.6¢/kg

Source: United States Embassy, London

The US Harmonised Tariff for Vegetables

In connection with the fruit and vegetable trade policy of the United States, reference is first made to vegetables and after that to fruit concerning the United States harmonised tariffs schedules (1990). The harmonised tariff schedules substantially define US fruit and vegetable trade policy, since the USA does not operate a policy of quotas or one of reference or minimum prices for fruit and vegetable commodities. Common to both the fruit and vegetable tariffs schedules are three separate tariff rates. Column 1 of the Harmonised Tariff divides into two sub-columns "General" and "Special". The rates of duty in the "General" column apply to all products of all countries which are not designated under the general rate as column 2 countries, and to all imports which are not eligible for a rate of duty provided in the "Special" column. Column 2 applies mainly to East European countries, the USSR, a number of Pacific countries and Cuba.

Symbols relating to the selected fruit and vegetable tariffs of the USA given in Tables 5 and 6 have the following meanings:

A denotes the Generalised System of Preferences established by UNCTAD

CA denotes the Canada Free Trade Agreement

E denotes the Caribbean Basin Recovery Act

IL denotes the United States/Israel Free Trade Area

The above are 'Special' column designations eligible for special tariff treatment. A large number of developing countries are known in the schedule as Beneficial Developing Countries. Direct imports from these countries to the United States are eligible for GSP treatment under UNCTAD. Independent Developing Countries qualifying number 101, and non-Independent Developing Countries number 27. Developing countries in the Caribbean are given special tariff treatment under the Caribbean Basin Recovery Act. This means that the United States fruit and vegetable trade policy is more liberal towards developing countries than the policy of the EEC. However, an examination of the harmonised tariff for fruit and vegetables reveals that 59 per cent of the vegetables, but only 27 per cent of the fruit fall into Category A, the Generalised System of Preference, and all such commodities are designated duty free. This finding further suggests that the harmonised tariff policy of the United States is more liberal, so less protectionist, than that of the EEC.

The US tariffs for fruit and vegetables, like those of Britain prior to 1973 which gave Commonwealth preference, and those of the EEC, are frequently scaled so as to give more protection to the domestic producer in his own marketing season.[11] The rate of duty is given in cents per kg and for certain vegetables as a per cent of the value. Per cent duty is frequently 50 or 25 per cent which penalises the exporter to a great extent. Rates per kg are very much lower in incidence than the rates in per cent. Looking at the harmonised tariffs for vegetables, for instance, tomatoes, the general rate is 4.5¢ per kg in the season from March 1 to July 14 and from September 1 to November 14, but it is 3.3¢ per kg at other seasons. This stepping of duty in relation to the domestic supply relates to a number of other vegetables, including lettuce and cucumbers. In the special category there is no tariff on tomatoes for specified countries. For tomatoes again, other countries in the special category are given a tariff of one cent less than those in the general category. Countries in column 2 face a 6.5¢ per kg tariff throughout the year, two cents higher than the general class. An important point, with great significance to developing countries, is the generally substantially lower rate on fruit and vegetables of the United States as compared with the European Community. At the same time, there are quite a number of important exceptions to this. These exceptions relate to commodities for which there is a percent rate at the high level of 25, 35 or 50 per cent of the value of the vegetables concerned. The US tariff has such a low incidence for many commodities compared with that of the EEC, and the United States does not apply quotas or minimum import prices. This means that the degree

---

[11] Acknowledgement is made to the US Embassy, London

of protection of the fruit and vegetable sector in the United States, is overall lower than the formidable package of measures which provide protection for this sector in the EEC. As stressed above, the US schedule, with a percent duty, represents formidable protection for producers of the commodities concerned. Moreover, these tariffs are chiefly year-round.

To give some further examples, onions have a year round tariff and the rate of duty 2 countries for onions is 5.5¢ per kg as opposed to the general rate of 1.3¢ per kg. The general duty on leeks is 25 per cent and the rate of duty 2 is 50 per cent. Cabbages and cauliflowers have a general rate of 5.5 per cent from June 5 to October 15. Countries in the rate of duty 2 category pay 50 per cent, a very high rate throughout the year. At other times of the year cabbages and cauliflower have a 12.5 per cent general duty. The general duty is increased when cabbage and cauliflower are cut or reduced in size, but in this case the rate of duty 2 is reduced from 50 per cent to 35 per cent.

The other main salads, lettuce and cucumbers are also, like tomatoes, treated as being sensitive to the American producers' incomes . Lettuce entering the United States from June 1 to October 31 must pay a general duty of 0.88¢ per kg and the rate of duty 2 is 4.4¢ per kg throughout the year. For other periods of the year the general rate of duty is 4.4¢ per kg the same rate as that for rate of duty 2 countries. More protection is given for cucumbers which are considered to be a sensitive commodity. From December 1 to the last day of February the general rate f or cucumbers is 4.9¢ per kg. From March 1 to April 30 it rises to 6.6¢ per kg. From May 1 to June 30 and from September 1 to November 30 it stays at this level and it falls to 3.3¢ per kg from July 1 to August 31. For countries in the special duty category, where the duty is applied, it is at lower rates over the same periods as the general duty. Countries in rate of duty 2 category all pay a higher rate, 6.5¢ per kg, and this is paid over all periods of the year. It is clear that the exporting countries most highly penalised and over the full year, are those countries in the rate of duty 2 class. Those countries in the special class, such as Israel and Canada are the most favoured. Such countries in the special class export to the United States either free of duty or at lower rates of duty than those in the general class countries, which gives countries in the special class a big advantage whether their commodities are free of duty or not, over all other countries. In fact, and it is of some interest, Canada, in the special duty class, is usually rated for duty, while some countries are free of duty. Countries in the general class have a more advantageous position than those many countries in the category rate of duty class 2. Vegetables, where a 50-25 per cent rate is applied are associated with very high rates of duty which apply to category 2 countries and over a large range commodities. In such cases category 2 countries are highly penalised.

Looking at the high percentage tariff for vegetables in more detail one finds that cabbages, Brussels sprouts and cauliflowers have a 50 or a 35 per cent tariff for class 2. Frequently, among these commodities too, the general rate is half of the 50 per cent or 35 per cent as the case may be. Leeks and other allicaceous vegetables have a rate of 50 per cent for Class 2 and a 25 per cent general rate. Carrots have a 35 per cent rate for Class 2 and a 17.5 per cent general rate while radishes attract a 50 per cent for Class 2 rate with a general rate of 17.5 per cent. Globe artichokes attract a 50 per cent Class 2 rate and a 25 per cent general rate. This also applies to asparagus throughout the season for Class 2, but for a certain period of the year the general rate is 25 per cent and only five per cent over another period. Celery has a 35 per cent rate for Class 2 and a 17.5 per cent rate of general duty. Mushrooms have a compound rate of 22 cents per kg and 45 per cent for Class 2, and a 11 cents per kg plus 25 per cent general tariff. Spinach attracts a Class 2 rate of 50 per cent and a general rate of 25 per cent. The group of vegetables including pumpkins attract a Class 2 rate of 50 per cent and a general rate of 25 per cent. This is repeated for okra which also illustrates what is common, for a special rate of duty, free to some countries and 20 per cent to others. Tomatoes have a high per kg Class 2 duty of 6.6 cents with the general rate at half this amount. For a number of other vegetables, Class 2 pay a 50 or a 35 per cent rate and general rate countries pay half this. It is quite clear that the general tenor of the tariff schedule of the USA is designed to protect US growers of vegetables from competition from overseas suppliers, especially in the home marketing season and to give a preferential tariff or a special rate of duty to a very few countries. The UK tariff schedule, which was operated from 1932 to 1973, when Britain joined the Community, which included preference to Commonwealth countries was very similar to the US Tariff schedule with some significant differences in tariff incidence. The general incidence of the UK tariff was higher overall than the US tariff. With the fruit and vegetable policy of the US, which applies today, a great number of developing countries qualify for the General System of Preference of the GATT, so developing country producers of vegetables pay lower duties than vegetable producers pay to the EEC, with the exception of ACP countries and member states overseas territories.

Table 6. The Harmonised Tariff Schedule of the United States for Selected Fruit, 1990

Commodity	Units	General Duty	Special Duty	Rate of Duty 2
Bananas	kg	Free		Free
Dates	kg	35%	Free E,IL; 9.9¢/kg CA	35%
Pineapples in bulk	kg	0.64¢/kg	Free E,IL; 5.9¢/kg CA	0.64¢/kg
Guavas and mangoes - entering Sep 1 to May 31	kg	8.27¢/kg	Free A,E,IL; 6.3¢/kg CA	33.1¢/kg
- entering at any other time	kg	8.27¢/kg	Free A,E,IL; 6.6¢/kg CA	33.1¢/kg
Oranges	kg	2.2¢/kg	Free E,IL; 1.3¢/kg CA	2.2¢/kg
Grapefruit - entering Aug 1 to Sep 30	kg	2.2¢/kg	Free E; 0.7¢/kg IL; 1.3¢/kg CA	3.3¢/kg
- entering in Oct	kg	1.8¢/kg	Free E; 0.5¢/kg IL; 1.0¢/kg CA	3.3¢/kg
- entering any other time	kg	2.9¢/kg	Free E; 0.9¢/kg IL; 1.7¢/kg CA	3.3¢/kg
Grapes - entering Feb 15 to Mar 31	m³ t	$1.41/m³	Free E,IL;$1.28/m³ CA	$8.83m³
- entering Apr 1 to Jun 30	m³ t	Free		$8.83m³
- entering at any other time	m³ t	$2.12/m³	Free E,IL;$1.696/m³ CA	$8.83m³

Source: United States Embassy, London

The Harmonised Tariffs for Fruit

One immediate difference between the harmonised tariff of the United States and the Common Customs tariff of the European Community is that the United States avoids placing tariffs on fruit which she does not grow. For instance neither the EEC of the USA grow bananas. Bananas in the tariff schedule of the United States enter free of duty from all countries, whereas a duty of 20 per cent is levied on bananas by the EEC. Citrus fruit, for the United States, is a very different matter because the USA is a considerable grower and trader in citrus[12]. For citrus, however, the United States does not penalise third countries as shown in the rates of duty on vegetables. Oranges are given a low general duty of 2.2¢ per kg and class 2 duty in the same amount. The special duty is 1.3¢ per kg or duty free. The same applies to all other oranges, mandarins, tangerines and satsumas, and clementines and Wilkins, for which there is no seasonal scale. The general rate for lemons is 2.75¢ per kg but double this amount for class 2. As for vegetables, for fruit generally there is no seasonal scale of duty. There are however, so far as fruit is concerned, a number of important exceptions, some of which are referred to in the following. Limes are treated very similarly to lemons. The rate of duty is 2.2¢ per kg and the class 2 duty is precisely double this. Grapefruit is given more protection though the rates of duty remain low. From August 1 to September 30 the general rate is 2.2¢ per kg and the class 2 rate is 3.3¢ per kg. Grapefruit entering the USA for the month of October is given lower tariffs, 1.8¢ general and 3.3¢ class 2. At any other time of the year the levy is more, 2.9¢ per kg general duty and 3.30¢ per kg class 2. Citrus fruit, which includes kumquats and bergamots, is curiously levied a 0.9¢ per kg duty and a substantial class 2 duty of 35 per cent. Grapes are given formidable tariff protection. From February 15 to March 31 the general duty is \$1.41 per m^3 but special duty countries are free (E, IL) and Canada pays \$1.28m^3. From April 1st to June 30th general duty payers are exempt and special duty countries, except Canada, are also free from duty. The duty for class 2 remains at \$8.83 per m^3 throughout the year but for general duty countries, the duty is 2.12¢ per kg from 1 July to 14 February.

All melons and pawpaws give rise to a class 2 tariff of 35 per cent. The general tariff for melons is also high. For instance, cantaloup melons, entering between August 1 and September 15, attract a general rate of 20 per cent and a class 2 rate of 35 per cent. Cantaloup melons entering at any other time attract a duty of 35 per cent at both the general and the class 2 rate. Water melons qualify for a 20 per cent general duty throughout the year. Ogden and galia melons attract a general duty of 3.5 per cent from December 1 to May 31 and 14 per cent at other times of the year. Other melons imported from December 1 to May 31 attract a general duty of 8.5 per cent and one of 35 per cent at any other time. Apples and pears are given only nominal rates of duty, if any. There is no general rate for apples, pears or quinces and the class 2 rate is only 1.1¢ per kg. Apricots attract a low general duty of 0.4¢ per kg as do peaches, but cherries and plums are free of general duty, except for plums entering from June 1 to December 31 when the general duty is 1.1¢ per kg. Rate of duty 2 is generally low for stone fruit, at 1.4¢ per kg on all stone fruit except cherries, which attract a high rate of duty of 4.4¢ per kg.

Strawberries are given a lower general duty of 0.4¢ per kg from June 15 to September 15 and a duty of 1.7¢ per kg for the rest of the year. The special rate is as usual a generous concession to the countries concerned, and the rate for class 2 is higher at 2.8¢ per kg. Raspberries and loganberries from September 1 to June 30 attract a general duty of 0.7¢ per kg and one of 2.8¢ per kg for class 2. The 2.8¢ per kg higher rate is applied to class 2 countries for all the berried fruit commodities other than the few which attract a duty for class 2 of 35 per cent. All berried fruit except raspberries are free of general duty and the few berries just referred to attract a tariff of 3.4 per cent. The harmonised US tariff for fruit shares in common with that for vegetables a low incidence of duty overall but a high percent duty for a few commodities. The percent tariff is applied to only a few fresh fruit commodities compared to vegetables, and is applied at 35 per cent and not 50 per cent which is common in the vegetable schedule. These are discussed later.

Oranges have a low rate of duty of 2.2¢ per kg, class 2 countries have the same rate. In the special column E and IL are free of duty, with Canada attracting a duty of 1.3¢ per kg. Rates in the three tariff classes are the same for the mandarin group. Lemons incur a general rate of 2.75¢ and a column 2 rate double that figure. There seems to be no particular consistency in the tariff rates. They often differ from each other by only a cent or so. In many cases, the class 2 rate may be the same as the general rate or double it. General rates of duty are mostly around 2.4¢ per kg, and an even lower rate, a figure around 0.4¢ per kg, is common. Avocados have a general rate of 13.2¢ per kg, the special rate varies from nil for Spain and 4¢ per kg for Israel, 10.4¢ for Canada and 33.1¢ per kg for all countries in class 2. The tariff on figs, with a general rate of 9.9¢ per kg has a similar structure to that for all the general tariffs on grapes. So there are examples of higher rates in cents per kg for fruit, but

---

12 European Community's Horticultural Trade, Moulton, Kirby. US Economic Research Service 1983.

it is not obvious why the last commodities should qualify for special treatment among the many others with low rates. Grapefruit has seasonally stepped low rates. A number of fruits, however, are given a percent rate, which as with vegetables, has a much greater incidence as a tariff than those fruits given a per kilo rate.

In summary, the United States tariff on fruit and vegetables, overall, is rather less protective than that of the EEC, and special provision is made for a large number of developing countries, which are either able to export to the United States duty free, or on the terms of the GSP.

The Marketing Order Programme of the United States

One important arm of policy for fruit and vegetables in the United States is the marketing order programme which operates at both the State and Federal level. A review of The Marketing Order Programme was given at the Warsaw Symposium in 1987.[13] Marketing orders are legally binding market plans which permit producers of fruit and vegetables in designated regions to control quantities and qualities of produce marketed and to engage in market support activities. Marketing orders are marketing plans designed by producers and the trade to achieve improved returns by establishing more orderly marketing conditions.

Federal marketing orders affect over half the tree fruits and 15 per cent of the vegetables in the United States. The farm gate value of the 33 crops concerned was $56 billion in 1984. In California, the leading area for fruit and vegetables, commodities affected by marketing orders have accounted for more then seventy per cent of the farm gate value of these crops.[14] In all about ninety different programmes are authorised under federal and state legislation. Marketing orders were established in the 1930's to benefit producers during the depression. Voluntary associations of producers were a prelude to legislation. Once government sanction is given on a State or Federal basis, all members of the production group must comply, hence the term "order". State orders only apply within a State. Marketing orders present three kinds of marketing, volume management, quality control and market support. The government gave producers monopolistic powers in the belief that uncontrolled competitive markets did not work well. Some price rises through the scheme were regarded as socially desirable, provided they did not lead to excess profits. More orderly marketing was seen as being consistent with the public interest in the longer term, even if it departed from free market conditions. Marketing orders have been controversial for years because of their monopolistic character and because it has not been demonstrated just how market order pricing differs from that in the free market. The Marketing Order Programme is just one more example, and the United States is far from being alone, of government sympathy towards agricultural producers. Policy is motivated by politics or by social considerations rather than economics. Marketing orders are often introduced at the expense of the tax payer who may see rises in price as an outcome.

Marketing orders sometimes have unfortunate effects. The effects of citrus fruit allocations, have been to maintain persistent differences in prices in the fresh and processed markets, especially for lemons. Consumers of processed lemon products have benefited at the expense of users of fresh lemons. A study made in 1987 suggests that the public interest would be better served by less restricted allocations in the fresh market.[15] Consumer groups in the States have been actively concerned about the programmes. Consumer groups in the EEC have yet to bring pressure on the CAP for fruit and vegetables. So far as is known, only consumers and the trade in West Germany have been critical of the EEC règime for fruit and vegetables. Market support programmes appear to have provided benefits of better organisation and improved information, and it is considered, that despite the acknowledged criticism of consumers, that the programmes have few adverse effects for them.

---

[13] Fruit and Vegetable Marketing Orders in the United States, 1937-1987, a Review, French B.C. University of California. Acta Horticulturae No 223 pp 48-57 1988.

[14] Marketing Order Alternatives. Use and Importance in Calfornia. French, Ben C, Tamini, Niniv and Nonkton, Frank. 1949-1975. Division of Agricultural Science Bulletin 1890 University of California Berkeley 1978.

[15] An Analysis of Economic Adjustments in the California - Arizona Lemon Industry. Kinney, William, Green, Richard and O'Connell, John. Giannini Foundation Research Report No. 337 University of California, 1987.

Voluntary management programmes are not likely to be expanded and a number of them have been terminated but quality control and market support programmes seem likely to be maintained. The use of marketing orders has now greater emphasis on quality controls and on market support programmes.

There are very few assessments of the dollar value of economic benefits and costs resulting from marketing orders. It is the opinion of French, that with one or two exceptions, if it would be possible to measure these volumes, we would discover that costs have been relatively small. Certainly they are small relative to the Federal support programme for dairy products, grains and field crops. The EEC budget for fruit and vegetables is also seen in this light to be much less expansive, for instance, than in the dairy sector. The EEC budget for support in the fruit and vegetable sector has been some four to five per cent or less, depending on the season, of total FEOGA budget expenditure. The fruit and vegetable marketing system in the United States is seen to be highly competitive even with marketing order intervention. It is claimed that a market may remain "highly competitive" even if the otherwise free market is interfered with to a lesser degree. It is also true of course that the United States market support on the fruit and vegetable sector is not part of a network of trade barriers of the level of that of the EEC.

That the fruit and vegetable marketing system of the USA is seen to remain highly competitive even with marketing order interventions is an important conclusion. Competition in the EEC fruit and vegetable trade, for the most part, concerns imports from third countries. This is reduced by the blanket of EEC trade regulations which operate against third countries. Moreover the value of fruit and vegetables imported by the EEC is significantly greater than that of the United States and a larger proportion of the trade is affected by the EEC regulations.

Producers of fruit and vegetables in the United States are better served by government, in so far as the United States provide their producers with a regular Situation and Outlook Service. This information has the effect that by improved market knowledge, the business risk of the producer is reduced, because producers of a significant proportion of crops can make sound market-oriented cropping decisions. A policy which gives a modest amount of market support for perennial crops, and a market outlook service for other crops would seem to be an advantage over the high protection of the EEC. This feature of policy to provide Situation and Outlook information for producers of fruit and vegetables in the EEC would have the have the effect of reducing expenditure on market support. Such a supplement to policy should be given serious attention by the EEC.

The Fruit and Vegetable Policy of Japan

The Japanese tariff schedule for certain fruit and vegetables is given in Table 7. The rate of duty column comprises four sub-columns, giving "General", "GATT", "Preferential" and "Temporary" rates. The General rates are based on the provisions of Article 3 of the Customs Tariff Law, and the Tariff Schedules appended are described in the complete schedule. The GATT rates are based on the General Protocol to the GATT in (1983), the provisions of Article 3 of the Customs Law and Article 5 of the Law are described. Preferential rates are based on the provisions of Article 8 of the Temporary Tariff Measures Law and Annexes 2, 3 and 4. Temporary rates are based on the provisions of Article 2 and Articles 6 to 8 of the Temporary Tariff Measures Law. Temporary rates are applicable to produce imported from April 1 1990 to March 1 1991. The Japanese fruit and vegetable trade policy is substantially described by the Japanese Customs Tariff for Fruit and Vegetables, and quotas still apply to a number of fruits[16].

Japanese Tariffs on Vegetables

The Japanese Customs Tariffs for Fruit and Vegetables includes general rates of duty of ten per cent, GATT rates and Temporary rates where applicable of five per cent apply to a wide range of vegetables, while a higher rate of duty of 25 per cent (General), 12.5 per cent (GATT) and 12.5 per

16 Acknowledgement is made to the Japanese Trade Office, London.

cent (Temporary), applies to sweetcorn. All vegetable tariffs apply for the whole year. They are not scaled according to the home season marketing for any commodity, unlike tariffs on certain commodities in the US and in the EEC tariff lists. Onions and the higher priced salad vegetables, and commodities like aubergines, mushrooms and bamboo shoots incur a general rate of duty of 15 per cent, and other types of duty are almost absent for these commodities. A few vegetable commodities are given a low rate in yen per kg. Cassava is given the higher rate of duty of 25 per cent (General). The characteristic of the Japanese customs duty for vegetables is that the incidence of duty is low. This applies to an overwhelming proportion of Japanese vegetable commodities.

Table 7. The Japanese Fruit and Vegetable Tariff - Selected Fruit and Vegetables, 1990

	Rate of Duty - per cent			
Commodity	General	GATT	Preferential	Temporary
Tomatoes	10	5		
Cucumbers	10	5		
Onions	10			
Cauliflower	10	5		5
Lettuce	10	5		5
Mushrooms	10	5		5
Sweetcorn	25	12.5		12.5
Potatoes	10	5		
Aubergines	10	5		5
Bananas - imported				
1 Apr to 30 Sep	20	20		
1 Oct to 31 Mar	40	40		
Pimentos	10	5		5
Pineapples	20			
Avocados	20	6	4	
Grapes	20			
Guavas, mangoes	20	6		6
Melons	20			10
Pawpaws	20	4	3	4
Apples	20			
Pears	20	8		
Apricots	20	10		10
Strawberries	20	10		
Raspberries	20	10		

Source: Japanese Embassy, London

Japanese Tariffs on Fruit

For fruit, the duty in general is twice that for vegetables, that is some 20 per cent, and GATT rates may be half this, or sometimes less. Certain preferential rates are nil and many are at a level similar to the GATT, and some are less than the GATT rate. Temporary rates of duty are 40 to 50 per cent in a number of cases and they are frequently 20 per cent or more, but there is a large range below 20 per cent, and for some commodities the temporary rate is nil. It is clear that the protection given to fruit is much greater than that for vegetables. Also for a number of fruits the tariff rates are stepped according to the home marketing season. Examples of the Japanese fruit trade policy are as follows. Bananas attract a general and a GATT rate of 20 per cent when imported from April to September inclusive, and rates of 40 per cent in each case, when imported from October to May inclusive. Avocados attract a 20 per cent general rate of duty which is reduced to six per cent under the GATT and to four per cent applying to countries with the preferential status (one country is free of duty) and the temporary rate is six per cent.

Oranges attract a general duty of 20 per cent, from June to November and a higher general duty of 40 per cent from December to May inclusive. There are no concessions on oranges under the GATT or to countries in the preferential and temporary categories. Lemons attract a general duty of 20 per cent, for the GATT it is five per cent and one temporary duty country is free of tariff. Grapes attract a general rate of duty of 20 per cent, 13 per cent under the GATT and 13 per cent under the temporary class when imported from November to February inclusive, but grapes are free of tariff from March to October inclusive. Apples attract a general rate of duty of 20 per cent which applies throughout the duty classes. The general tariff for pears is also at 20 per cent, but the GATT concession is eight per cent. Rates for apples and pears, like a number of other fruits, are applied for the year with no seasonal differentiation. Also, as with a number of vegetables with a ten per cent duty, a number of fruits with a 20 per cent general duty show concessions to the GATT at half the general rate.

It is clear that the Japanese fruit and vegetable trade policy creates a substantial barrier to fruit from developing countries and for certain fruits mentioned above it is a formidable barrier. Tariffs on vegetables are less than those for fruit, but the Japanese vegetable tariffs also interfere with the export trade of developing countries, though to a lesser extent than the tariffs on fruit.

The Consequences of UK Accession to the European Community for the UK Market for Apples.

One of the consequences of EEC membership are the dramatic results stemming from the switch in policy in the UK from one of protection and Commonwealth preference to one of free trade with EEC partners. On UK accession to the EEC, France, like all member states now had a free market for fruit and vegetables in Britain. The increased competitiveness for France in the UK apple market had crucial and damaging results both for the sector in Britain and for third country suppliers to Britain. The apple sector in Britain found it could not compete with France, now on equal terms, and the third country exporters were obliged to cope with the protection in the British market due to the EEC fruit and vegetable règime being applied to Britain.

The effects for the UK apple market are well demonstrated by Atkin and Blandford who used a Markov model applied to UK apple imports over the period 1963-79. They found the fit of their model indicated that changes in market share have been systematic, stable and of long duration.[17] The estimated transitional probability matrix shed light on the relative competitive strength of the different exporting countries. The large increase in the French market was examined and an estimate derived of the effect of UK membership of the EEC. For the years after which transition was complete, it was estimated that UK membership increased the French market by more than 26 per cent and decreased the share of Australia and South Africa by 18 per cent and ten per cent respectively.

UK imports are an important segment of international trade in apples. During 1971-7 this averaged 360 thousand tonnes or 12 per cent of the world total. Despite a doubling of world trade in apples from new import markets such as Hong Kong, Saudi Arabia, Taiwan and the USSR in the 60's and 70's, the UK's share of world exports has remained constantly in the 10-15 per cent range.

---

[17] Structural changes in market shares for apples in the United Kingdom. European Review of Agricultural Economics. Atkin, Michael and Blandford, David pp 313-326 1982

Table 8. Trade share data used in estimation - per cent

	Australia	Canada	France	South Africa	United States	Rest of the World
1963	36.5	10.6	2.0	20.3	8.1	22.5
1964	30.6	10.2	4.9	21.1	11.2	22.0
1965	33.8	10.1	5.2	20.6	10.9	19.4
1966	26.7	7.4	9.7	26.2	9.8	20.2
1967	21.0	5.4	22.7	22.2	7.1	21.6
1968	19.7	4.1	23.0	24.2	4.7	24.3
1969	24.5	3.7	27.1	25.3	1.1	18.3
1970	26.8	2.5	22.8	25.2	2.0	20.7
1971	25.9	1.7	26.5	23.3	1.6	21.0
1972	19.5	1.3	28.7	27.7	1.9	20.9
1973	14.6	1.6	42.4	20.4	1.9	19.1
1974	10.4	0.3	48.8	19.2	0.7	20.6
1975	9.0	0.4	51.6	16.9	2.0	20.1
1976	6.6	0.3	55.4	16.1	0.9	20.7
1977	4.1	0.8	57.1	10.2	1.3	26.3
1978	2.4	0.7	55.7	16.7	2.6	21.9
1979	3.8	0.4	65.9	16.8	2.3	10.8

Source 1963-77 UN Trade Year book 1978-9 Fruit and Tropical Products Year book. Commonwealth Secretariat

Table 8 shows that French competition in the UK apple market before and after UK accession to the Community on January 1st 1973. French competition had already grown markedly from small beginnings in 1963, when France only had two per cent of the market share to a 28.7 per cent share in 1972, the last year before the UK acceded to the Community. The 1972 pattern of the per cent share of apple imports to the UK shows that, to that time, imports from France increased at the expense of Australia, Canada and the United States. Australian imports had a market share of 36.5 per cent in 1963 to fall to 19.5 per cent in 1972. The Canadian market share was 10.6 per cent in 1963 and fell to 1.3 per cent in 1972. The United States market share fell from 8.1 per cent in 1963 to a 1.9 per cent share in 1972. Over the period 1963 to 1972 the South African share actually increased from 20.3 to 27.7 per cent. From 1972 to 1979, France more than doubled her market share to reach a 65.9 per cent market share in 1979, made possible because of the free market in the EEC which gave France more competitive power. By 1979 Australian exports to Britain dwindled to give her a 3.8 per cent share, and Canadian sendings fell to a 0.4 market share. The United States share was further reduced to 2.3 per cent, and South Africa with a 27.7 per cent market share in 1972, declined to a 16.8 per cent share in 1979. Meanwhile the area of dessert apples grown in Britain fell from 232,580 ha in 1967 to 139,880 ha in 1988, a decline of 40 per cent, a crushing blow to the dessert apple sector in Britain.

France gained a larger market in the UK based on her low cost per unit apple crop - Golden Delicious, which yields twice that of English dessert apples which are grown in a less favourable climate. Also, most consumers have a preference for it. At the same time Golden Delicious is a superb supermarket commodity, because if its consistent quality and size. Added to this are the advances in storage over the years which means that Golden Delicious can now be marketed every week of the year.

The French advantage is the comparative advantage in climate which free trade within the EEC unleashed. In 1947 the dessert apple industry in the UK was given a policy of protection by tariff and quota limits on dollar fruit by a generous government. This policy had to go on UK accession to the Community. It is true that by 1967 before the accession of the UK the French had already gained a 22.7 per cent market share and the area of dessert apples in the UK had already begun to fall. The results of the new French competition in the UK apple market by 1979, not only dramatically reduced the market share of Canada and the United States but also that of southern hemisphere apples, and the market share of South Africa less so, but it crushed the UK domestic apple sector. The UK is on the climatic margin for optimal production of apples. Formerly this industry was maintained in profitability by protection. Now that the UK has a strong competitor in the free market of the EEC comparative advantage has come into this own.

Voluntary Import Restraints in the Apple Market of the EEC

There is one aspect of more recent EEC policy which merits attention in some detail. This is the EEC response to apple imports from third countries in the southern hemisphere. Fifty eight per cent of all southern hemisphere exports of fruit to the EEC are apples and pears, mainly apples. For decades now the EEC has witnessed structural surpluses of apples and pears which Brussels has not been able to contain.[18] Moreover it is contended that the intervention price system has actually encouraged production such that producers in France and Italy have been growing for intervention. This is the firm contention of the writer and it is supported by Avermuete, of the Belgian National Farmers Union, who suggests that the intervention system is the likely cause of the recurrent surpluses of apples and pears in the EEC.[19]

Professor Winter, Director of the Bavendorf Research Institute in Bavaria, an organisation which monitors apples and pear yields throughout the European Community, has recently stated that the surpluses of apples and pears are still with us and that they will continue in the future as well as the tendency of increasing imports of "off-season" apples from the southern hemisphere.[20] The main cultivar in the EEC is Golden Delicious, but because of the age structure it will decrease in production in Europe in the future. The main producer of Golden Delicious is France, where in the south it is 60 to 75 per cent of the total production. The second cultivar in Europe and the world over is Red Delicious which has first place in the United States. Granny Smith is mainly grown in countries in the southern hemisphere but it is also grown in France and Italy. In Central and Northern Europe some old varieties such as Boskoop, Ingrid Marie and Bramley, a cooking apple in the UK, have a share, but newcomers such as Jonagold, Gloster, Idared and Elstar are increasing.

Technical improvements in storage have played an important part in the market for apples in the EEC. Over the last twenty years apple storage has so improved that the dessert variety "Golden Delicious", for instance, as mentioned before, can be marketed from store throughout the year instead of six months as formerly. Apples from the southern hemisphere, freshly picked, are sent to Europe in the spring and early summer. The arrival of southern hemisphere apples now coincides with a substantial proportion of the Community's domestic supply. In the late 1970's and to a lesser extent in 1981 and 1982, countervailing charges were frequently triggered in the UK and Dutch markets by supplies from Chile, Argentina and New Zealand.[21] Southern hemisphere countries supplying the European market with apples are South Africa, Argentina, Chile, New Zealand and Australia. Because of more competition in the EEC arising from southern hemisphere and domestic supplies coming in the market together, COPA, the Confederation of Farm Unions in the EEC, pressed Brussels to limit imports from the southern hemisphere in order to raise the price of EEC apples in the spring and early summer, so as to offset the price fall which now arose towards the end of the EEC apple market season. Brussels responded with sympathy and negotiated a series of "gentlemen's agreements" known as Voluntary Import Restraints, with countries in the southern hemisphere. The southern hemisphere countries had no choice but to agree to restrict their supply of apples in certain seasons

---

18 EEC Surpluses of Apples and Pears, a Perspective. Hinton Lynn Acta Horticurae No 155 pp 29-32 1985

19 The EEC Market Policy for Fruit and Vegetables Avermuete, U. Acta Horticurae No 233 pp 37-40. 1988

20 Trends of Capacity of Apple Production Western Europe, North America and the Southern Hemisphere. Winter F, Acta Horticulturae No 233 pp 112 to 119, 1988.

21 The Impact of the EEC's Reference Price System on the Marketing of Fruit and Vegetables in the UK. Williams Helen, Ritson Christopher. Report 31 Vol 1 Main Report Department of Agricultural and Food Marketing. University of Newcastle upon Tyne. 1987.

because they believed without agreement a quota system would be imposed by Brussels on their apple exports. This was something they did not want because quotas would have meant an even more severe disruption of the apple market. The first of these agreements was established in 1976. Voluntary agreements were also reached in 1979, 1983, and 1988 seasons of high EEC crops.

In seasons when the voluntary agreements are enforced by Brussels, southern hemisphere countries are obliged to inform Brussels in advance of their marketing season of their likely sendings by variety and by country of destination. The apple trade with third countries was already regulated by a year-round reference system, a safeguard clause, quality norms and export subsidies. By itself the common customs tariff does not amount to a very significant trade barrier, and only rarely have countervailing levies been paid on southern hemisphere apple imports.[22] There is, of course, an intervention system for apples and pears in addition. What became the basis for the Brussels imposition of voluntary import restraints is the safeguard clause. Article XIX of the GATT sets out a procedure such that the contracting party may suspend its trade obligations in whole or in part when a product is imported into the EEC from that contracting party in such quantities as to cause, or threaten to cause, serious injury to domestic producers. This enabled the EEC to limit third country imports by quotas, additional levies, or to ban third country exports completely. Despite voluntary import restraints on apples, producer groups in the EEC still seek additional protection under the safeguard clause. While EEC policy for apples was at this time, wholly concerned to protect the producer, it is interesting to see the trade reaction to the diversion of Chilean imports from the market. The trade, of course, would like to see a free market in all fruit and vegetables because, as they see it, the present policy raises the price to the consumer and limits consumer choice. It is true that the restriction of the imports of southern hemisphere apples means that fewer southern hemisphere apples freshly picked are available in the spring and early summer.

The Trade Press, of course, sees Voluntary Import restraints as disturbing the market. It was reported that when Chilean apples were diverted from the market, a French Company had some 2,000 tonnes on the water when the Commission suddenly operated the apple licensing procedure. When the fruit arrived it was subsequently bonded and the Company resorted to the EEC court to get it freed. The application was allowed under a regulation which takes account of goods in transit deemed to be safe from sudden restrictive decisions. The ruling of the Court was that the Commissions decision relating to the ban of apples from Chile, only applied to the one Company. In the event many of the Chilean apples were released on to the market through the fact that "samples" could be sold. Consignment notes for thousands of samples of less than 20 cartons completely cleared one apple store.

The Confederation of Importers and Marketing Organisations declared in relation to the ban on Chilean apples, that it led to severe disturbances in the market with negative affects for consumers in the Community and severe financial losses for producing countries. Also at the same time, the CIMO went further and attacked the reference price system. "The Commission has disturbed the market almost every day in the course of the year with high supplementary levies, which made it impossible for the trade to supply consumers to their wishes of apples, peaches, plums, lemons, clementines, tomatoes and cucumbers. Such measures which block the free world trade in fresh produce, have led to a quality decrease which in turn leads to a decrease in per capita consumption. In the long term CIMO contended that the situation would affect EEC production. The first signs were being seen, they said. Finally, they said that the intervention of fruit and vegetables, most of which is destroyed, collectively amounts to 8 kg per head in the Community. Supplies of apples grown in the southern hemisphere for 1988 and 1989 are given in Table 9.

Table 9 shows imports of southern hemisphere apples for 1988 and 1989 which arrive over the second half of the EEC marketing season. Southern hemisphere apple imports were 16 per cent up in 1989 as opposed to 1988. They were appreciable up for Chile at 18 per cent. New Zealand exports were up 22 per cent and South Africa 17 per cent. It would seem from the data that New Zealand was a worse offender in 1989 than Chile in 1989.

---

[22] The Impact of Voluntary Export Restraints on the United Kingdom Import Apple Market Rae A.N. Acta Horticulturae No 223 pp 326-333 1988.

Table 9. Southern Hemisphere Apple Imports to the EEC - tonnes

	1988[a]	1989[b]	Per cent change
Argentine	70,000	72,000	+ 3
Australia	11,000	11,000	-
Chile	142,000	168,000	+18
New Zealand	115,000	140,000	+22
South Africa	166,000	194,000	+17
Total	**504,000**	**585,000**	**-16**

[a] 1988 quotas set by the EEC licence system
[b] Targets for voluntary self-limitation

Source: The Fruit Trades Journal 1990

Table 10. EEC Apple Production and third country Imports

	1984-5	1985-6	1986-7	1987-8	1988-9	1989-90
EEC production[a]	8370.6	7433.6	8260.3	7543.6	8747.6	7946.6
	1984	1985	1986	1987	1988	1989
Third Country						
Imports[b]	444.5	460.9	490.8	543.5	602.2	596.4
Exports[b]	189.5	188.5	200.5	184.4	190.4	160.7
Withdrawals[a]	660.9	184.6	354.1	438.4	660.1	596.4

Source: [a] Eurostat Cromers Dg V1) [b] Eurostat

Table 10 gives withdrawals of apples for marketing years 1984-5 to 1989-90. For the same period imports of apples from third countries are shown. The figure for imports may be taken as southern hemisphere imports since any imports from other third countries must be very small, if indeed there were any. Withdrawals were high in 1984-5 at 660.9 thousand tonnes, when the level of the EEC crop was also high. Withdrawals in 1986-7 were 354 thousand tonnes, when there was another high EEC crop. In 1987-8 and 1988-9 there were high withdrawals each year of 438.4 thousand tonnes and 660.1 thousand tonnes respectively. In 1988-9 withdrawals were the same as those in 1984-5.

The 1984-5 season witnessed the highest EEC production for all of the marketing years given. This suggests that the level of withdrawals in any year is a function of the level of production. The two years 1984-5 and 1988-9 were each years of peak withdrawals. It had been suggested to me that third country imports were the cause of high withdrawals. Clearly this is not so. When production is around 7.5 million tonnes as in 1985-6 and 1987-8, withdrawals are roughly half the peak level of the two seasons of 660 thousand tonnes. The third country imports of apples can hardly be seen to influence the amount of apples withdrawn. This is substantially a function of the level of EEC production. Imports from third countries are the highest at 602.2 thousand tonnes in 1988 and lowest at 444.5 thousand tonnes in 1984. The difference between the highest and the lowest imports is only 158 thousand tonnes, whereas a look at EEC production shows the difference between the extremes to be 1314 thousand tonnes.

The EEC Stabiliser Mechanism for Apples

An experimental scheme designed to stimulate apple prices and reduce the costs of EEC support in intervention payments was introduced in 1989 and adopted for 1990. An EEC spokesman said that the experiment had been positive and it had helped to limit the costs of the apple règime. This measure was designed to tackle the oversupply of apples which led to 354,000 tonnes being taken into intervention in 1986-7 and 591,000 tonnes being taken off the market in 1987-8, almost a tenth of EEC apple production. The method is to regulate apple size so that only apples of a certain size may be marketed. By increasing the size of the product that can be sold, less will go into intervention and there will be less chance of price cuts being triggered under the stabiliser règime and there will be less competition from smaller apples. This measure, like other measures to save on intervention funds, since the first one was introduced for tomatoes in 1987, is a complete reversal of the very high level of protection relating to apples. The stabiliser règime breaks the theme, so basic to the EEC fruit and vegetable règime, that of supporting the producer. This policy will also play a positive part in reducing the recurrent apple surpluses which the intervention system has been thought to encourage.

CHAPTER 4

THE WEST GERMAN MARKET FOR FRUIT AND VEGETABLES

The West German Fruit and Vegetable Economy

The fruit and vegetable imports to West Germany are not only the biggest in Europe but they are not matched for size by any other country including the United States. In 1987 the United States had a home production of vegetables of 10980 thousand tonnes and imports of 1680 thousand tonnes worth 721 million dollars. Home production of fruit was 28257 thousand tonnes supplemented by imports of 1190 thousand tonnes worth 659 million dollars. Imports of fruit and vegetables in the United States amounted to 1380 million dollars and those of West Germany to an equivalent of 1614 million dollars.[1]

In West Germany the area under vegetables is little more than the area under fruit, each being some fifty thousand ha. (This chapter relies on the valued publications of the Zentrale Markt und Preisberichtstelle, the West German data centre for agriculture, forestry and food.[2]) This is in marked contrast to the United Kingdom where the area under vegetables greatly exceeds that under fruit. Also the area of protected vegetable cultivation in West Germany is low; at some one hundred thousand ha. The relative dearth of field vegetables and of vegetables under glass in West Germany is the result of successful competition by the Netherlands in this sector over several generations. In a very real sense Holland is the market garden of West Germany, if not of Western Europe. As a result, West Germany is less self-sufficient for field vegetables and for vegetables grown under glass than she is for fruit. The level of vegetable production in West Germany runs at some 1,500 thousand tonnes or half that of fruit. In 1988 vegetable production by value was 1,324 million DM while fruit production by value was 2,097 million.

Imports of vegetables are 2,500 million DM a year (1988), nearly twice the value of West German domestic production. Imports of fruit run at between five and six billion DM or twice the value of home production. Imports of fruit of course include non-temperate fruit, which West Germany, like the United Kingdom, is obliged to import on a regular basis. Exports of both fruit and vegetables are low and they represent mainly border trade. Vegetable exports are some 200 million DM a year and fruit exports are twice this level. West Germany is one third self-sufficient for vegetables and from 40 to 60 per cent for fruit excluding citrus fruit depending on the level of home production in a given year.

Vegetables in West Germany - Area and Production

The West German production by area and volume is given in Table 1. The area of vegetables grown in West Germany, as already stated, is small, stemming from West Germany's heavy reliance on Holland for a large part of her supply and a yet larger part of her supply of salad vegetables.

---

[1] Data for the United States provided by Christine Chapman. US Embassy London.

[2] ZMP Bilanz - Gemüse. ZMP Bilanz - Obst.

Table 1. West Germany - Vegetable Production 1987, Areas and Quantities

Items as listed in Eurostat

	'000 ha	'000 tonnes
Cauliflowers	3.3	75.1
Brussels sprouts	0.4	5.5
Cabbage	5.4	286.1
Other brassicas	5.4	167.3
All brassicas	**14.5**	**534.0**
Celeriac and celery	1.4	41.7
Leeks	1.7	42.9
Lettuce	3.8	82.0
Endive	0.4	6.1
Spinach	2.3	33.6
Asparagus	4.3	12.2
Other vegetables (leafed or stalked)	0.8	12.1
Vegetables leafed or stalked	**14.7**	**230.6**
Tomatoes	0.3	18.5
Cucumbers	0.5	33.8
Gherkins	1.0	20.8
Cultivated for fruit	**1.8**	**73.1**
Kohl-rabi	1.6	40.1
Carrots	4.0	139.4
Onions	2.1	84.9
Beetroot	0.5	17.4
Radish	0.1	1.6
Other roots and tubers	1.1	6.5
Roots and tubers	**9.4**	**289.9**
Peas	1.7	8.1
French and runner beans	3.1	28.7
Other pulses	0.6	3.3
Pulses	**5.4**	**40.1**
Cultivated mushrooms	**-**	**37.0**
Other fresh vegetables	**3.5**	**77.2**
Total of vegetables specified	**49.3**	**1281.9**

Source: Eurostat 1989

The area of vegetables grown in West Germany is 49.3 thousand ha, and the total production is 1.3 million tonnes. Some fifteen per cent by area is in brassicae and a quite similar amount is in leafed or stalked vegetables. These together are three fifths of the total vegetable area. Vegetables cultivated for their fruit, substantially salads, are very low in area, only two thousand ha in the total vegetable area of fifty thousand ha, so dependent is West Germany on Holland for this class of vegetable. Roots and tubers are important with some twenty per cent of the total vegetable area. Pulses account for 5.4 thousand ha, a little over ten per cent of the total vegetable area, while "other fresh vegetables" comprise 3.5 thousand ha. The brassicae are more prominent in volume at 534 thousand tonnes, as are leafed and stalked vegetables at 230 thousand tonnes but the small area of vegetables grown for their fruit, influence volume more in proportion at 73 thousand tonnes, while roots and tubers have a production of 290 thousand tonnes. The production by volume of pulses is 40 thousand tonnes from an area of 5.4 thousand ha. Mushrooms account for almost three per cent of the total production of vegetables by volume, a big share. In the West German market most of the mushrooms are grown for canning or bottling, as opposed to the UK where 90 per cent by weight are put on the fresh market.

West German fruit production by area and volume is given in Table 2. Of the total of 55 thousand ha, half the area is in apples and pears, substantially apples, nearly a quarter is in stone fruit, mainly cherries and plums, and almost one fifth of the total fruit area is in soft fruit, of which strawberries are the most important. Production by volume of all fruit is 2.2 million tonnes, half of which is apples, and a quarter stone fruit. Soft fruit represents over one eighth of all fruit production.

The large proportion of fruit to vegetables in West Germany, not typical of countries in northern Europe, is accounted for by the division after World War II. Most of the fruit is in the Western part and most of the vegetables in the Eastern part.

Table 2. West Germany - Fruit Production 1987, Areas and Quantities

Items as listed in Eurostat

	'000 ha	'000 tonnes
Apples	26.6	1057.0
Pears	1.6	285.0
Apples and pears	**28.2**	**1342.0**
Peaches	0.1	23.4
Apricots		1.8
Cherries	8.2	232.9
Plums	3.8	321.8
Stone fruit	**12.1**	**579.9**
Black and red currants	1.2	130.2
Strawberries	7.7	53.5
Raspberries	0.1	25.7
Gooseberries	0.7	81.8
Soft Fruit	**9.7**	**291.2**
Other fruit	**5.5**	
Total of fruit specified	**55.5**	**2213.1**

Source: Eurostat 1989

Table 3. West German Trade in Vegetables, 1988 - tonnes

	Imports	Exports
Tomatoes	383067.7	2581.1
Onions	318869.1	13943.4
Shallots	3987.4	26.0
Garlic	7983.1	85.2
Leeks	30595.9	991.0
Cauliflower	141744.3	2361.2
Brussels sprouts	28882.4	35.3
White cabbage	18399.7	17118.7
Red cabbage	1949.9	1188.7
Savoy cabbage	6802.9	219.2
Kohl rabi	12790.7	337.5
Chinese cabbage	7997.7	3169.3
Other cabbage	4716.3	1637.3
Iceberg lettuce	28100.5	305.2
Lettuce	90596.9	352.3
Other salad	16482.2	573.3
Chicory	15101.3	64.9
Endive	40085.3	1024.8
Carrots	113620.4	4002.8
Celery	10189.3	521.1
Horseradish	2547.5	423.5
Other root vegetables	47234.5	3546.6
Cucumbers	273272.0	3806.8
Pickling cucumber	53533.7	1605.7
Small gherkins	3951.0	233.9
Peas	2006.1	2110.5
Beans	13929.9	5783.1
Other pulses	881.8	-
Artichokes	1902.5	21.0
Asparagus	26059.9	346.6
Aubergines	13599.4	42.4
Blanched celery	3445.1	72.0
Agaricus mushrooms	13779.6	713.4
Chanterelle mushrooms	2405.4	88.1
Yellow boletus mushrooms	166.2	21.0
Oyster mushrooms	299.6	47.0
Other mushrooms	154.1	21.2
Truffles	8.7	1.6
Sweet Peppers	148438.2	1046.1
Other peppers	4808.7	143.8
Spinach	4766.4	6462.4
Other leaf vegetables	2456.3	82.3
Mangolds and chard	682.2	-
Olives	200.1	-
Capers	76.7	1.8
Fennel	6857.0	27.6
Sweet corn	396.0	14.0
Courgettes	13243.6	79.7
Other vegetables	14195.7	1256.5
Total	**1937913.2**	**78496.9**

Source: ZMP Bilanz - Gemüse 1987-88

Table 4. West German Imports of Vegetables by Country of Origin - Quantity and Value 1987

Country	Quantity tonnes	Quantity per cent	Value '000 DM
France	149430	7.3	300989
Belgium/Luxembourg	101052	5.0	163666
Netherlands	827688	40.5	1382850
Italy	321966	15.8	357400
United Kingdom	5996	0.3	3265
Ireland	-	-	-
Denmark	2194	0.1	1333
Greece	50904	2.5	110901
Spain	312851	15.3	401948
Portugal	514	0.0	665
EEC	**1772595**	**86.8**	**2723917**
Yugoslavia	598	0.0	1810
Austria	16071	0.8	18842
Switzerland	-	-	-
Canary Islands	43538	2.1	60923
Turkey	18989	0.9	23056
Other W. European countries	-	-	-
West Europe	**79196**	**3.9**	**104631**
Albania	895	0.0	1150
Bulgaria	8467	0.4	6028
Czechoslovakia	9271	0.5	2087
Poland	38869	1.9	38722
Rumania	15558	0.8	11332
Hungary	44028	2.2	26686
USSR	-	-	-
East Europe	**117088**	**5.7**	**86005**
Ethiopia	558	0.0	1314
Kenya	644	0.0	1695
	207	0.0	529
Egypt	5982	0.3	5313
Morocco	13177	0.7	18190
Israel	7015	0.4	8696
South Africa	6349	0.3	4487
USA	17	0.0	104
Chile	3347	0.2	1764
New Zealand	3684	0.2	2168
Australia	21242	1.0	12248
Other overseas countries	2920	0.1	5271
Overseas countries	**65142**	**3.2**	**61779**
Unallocated quantity	7585	0.4	9481
Total	**2041606**	**100.0**	**2984913**

Source: ZMP Bilanz - Gemüse 1987-88

Figure 1. West German Imports of Vegetables by Country Of Origin

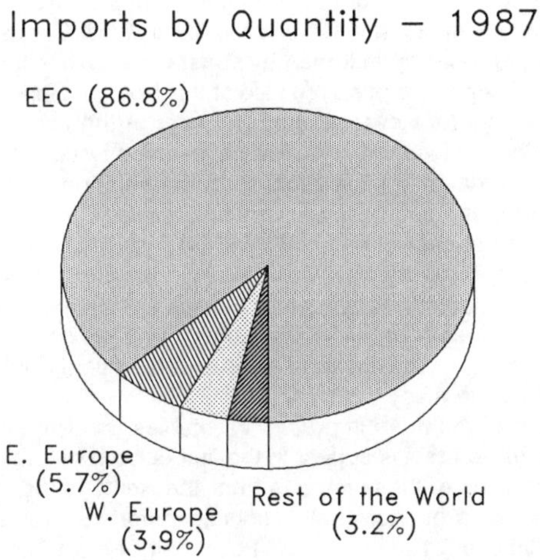

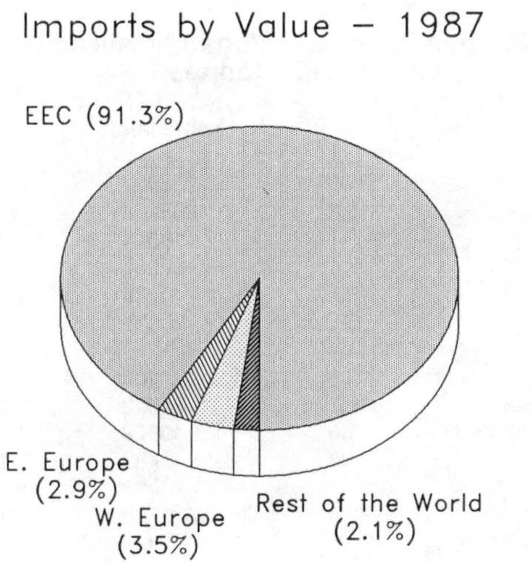

Source: ZMP Bilanz – Gemuse 1987–88

The West German Market for Vegetables

Imports and exports of vegetables in West Germany are given in Table 3. The table shows a substantial amount, nearly two million tonnes of imports over a considerable range of produce. Total exports of vegetables are only four per cent of imports. In quantity tomatoes stand out as the biggest import, even greater than the more bulky imports of onions. Other large imports by weight are cauliflower, carrots and peppers, but the range of different vegetables is impressive.

Imports of vegetables by country of origin is shown in Table 4 and in diagrammatic form in Figure 1. Eighty-seven per cent comes from other members of the EEC, forty per cent from Holland alone. Italy makes a relatively light contribution and the Italian-West German trade connection in fruit and vegetables is represented more by fruit than by vegetables. While Italian vegetable imports are 15.8 per cent of the imported supply, imports are also at this level for Spain. These two countries and the Netherlands together account for over seventy per cent of the imports of vegetables into West Germany. Imports from Belgium-Luxembourg at five per cent are twice those of Greece which has the potential to send more. Portugal in 1987 is seen not to be sending vegetables to West Germany. Imports from other West European countries are very low but the Canary Islands export of tomatoes in the winter shows up at 2.1 per cent of the total West German imports of vegetables. East Europe contributes 5.8 per cent of vegetable imports, most coming from Hungary which has a most successful specialist export agency. All countries in Eastern Europe are interested in the hard currency trade which West Germany offers. Those countries classified as 'overseas countries' only contribute 3.6 per cent. Much of this concerns out of season vegetables and the supply of 'exotics' to meet the rising demand of consumers in West Germany.

West German imports of the more important vegetables and from the three main sources are shown in Table 5. Imports of vegetables appear in the first column of Table 5 in descending order of import value, while the quantities of those imports from the Netherlands, Italy and France are given alongside. Tomatoes alone are a quarter of all vegetable imports by value, while cucumbers are 16 per cent, peppers 8.7 and lettuce 8.1 per cent. Well over half of West German imports of vegetables are represented by these four salad crops. Other high ranking imports are asparagus, onions, cauliflowers and carrots. The eleven named commodities account for 82.8 per cent by value of all vegetable imports and the remainder which are numerous, show great diversity and cover the whole spectrum of demand.

Table 5. West German Imports of Vegetables from Principal Sources, 1987 - tonnes

	Per cent[a]	Netherlands	Italy	France
Tomatoes	24.9	234007	14070	9404
Cucumbers	15.8	205164	971	2929
Peppers	8.7	18574	26065	2827
Lettuce	8.1	62267	7837	8718
Asparagus	6.6	4110	116	10723
Onions	5.6	120635	28528	12119
Cauliflower	4.4	1744	49689	53676
Carrots	2.6	24005	59293	12459
Other root vegetables	2.3	24072	21313	255
Pickling cucumbers	1.9	19072	7751	-
Other salad	1.7	483	12423	5119
Total	82.8	827620	321970	149433
Per cent total imports		40.5	15.8	7.3

[a] Per cent of imports by value

Source: ZMP Bilanz - Gemüse 1987-8

The Netherlands, as mentioned before, dominates the imports of vegetables to West Germany with 40.5 per cent. Holland supplies much the most of the tomatoes 234 thousand tonnes, as against 14 thousand from Italy and nine thousand from France. This is repeated with cucumbers where Holland supplies 205 thousand tonnes and Italy and France together, a minority of four thousand tonnes. For lettuce imports, Holland again supplies the lion's share. By contrast, the weight of imports of peppers comes from Italy, but again the Netherlands is a good second. Asparagus is substantially imported from France, nearly eleven thousand tonnes, with Holland supplying four thousand tonnes. The Netherlands is a minority supplier of cauliflowers, while Italy and France supply cauliflowers in almost equal amounts. This is a case where the maritime climate of Brittany and the mild winter of Italy unite to supply most of the cauliflowers, a vegetable sensitive to frost. Italy is the major supplier of carrots but a big proportion also come from Holland, while Holland is again supreme in the imports of onions. This is very much to Holland's credit with superb ripening and storage of the crop which comes to Europe in large quantities from more favourable (hotter) climates throughout the World. Overall Italy contributes 15.8 per cent of the vegetable imports and France 7.3 per cent.

Table 6. West German Vegetable Supply, Self-sufficiency and Consumption per Head, 1987-8

	'000 tonnes	Self-sufficiency per cent	Consumption kg/head
Large vegetables			
White cabbage	261.8	98.4	3.6
Red cabbage	78.6	95.3	1.2
Savoy cabbage	44.6	81.8	0.6
Green cabbage	13.7	100.0	0.2
Late kohl rabi	12.8	100.0	0.8
Late carrots	99.9	100.0	3.8
Celery	50.3	74.6	0.7
Leeks	73.2	52.7	1.0
Broad beans	3.0	100.0	0.1
Total Large Vegetables	**637.0**	90.0	
Fine vegetables			
Cauliflower	217.4	31.1	3.2
Brussels sprouts	40.0	12.4	0.6
Chinese cabbage	25.6	100.0	0.4
Spinach	30.1	100.3	0.6
Early kohl rabi	43.9	83.1	0.8
Early carrots	163.3	25.8	
Asparagus	37.8	29.1	1.2
Peas	6.9	105.8	1.2
Runner beans	33.1	78.0	2.1
Lettuce	200.0	36.9	2.9
Chicory	18.2		0.3
Onions	429.6	17.8	6.9
Tomatoes	430.8	3.9	14.7
Cucumber	305.3	10.0	4.3
Pickling cucumbers	88.1	21.2	1.4
Total Fine Vegetables	**2070.1**	21.9	
Sundry vegetables	492.7	26.3	13.6
Total All Vegetables	**3199.8**	36.2	

Source: ZMP Bilanz - Gemüse 1987-88

Vegetable production, self-sufficiency and consumption

West German production of vegetables is given in Table 6 classified as 'large vegetables' and 'fine vegetables'. Whereas the West German production shows a great variety of vegetables, West Germany is seen to be in conspicuously deficient in self-sufficiency for such crops as tomatoes, onions, asparagus and Brussels sprouts. Overall West Germany is only 36.2 per cent self-sufficient for vegetables. The degree of self-sufficiency, of course, mirrors the amount of imports. Table 6 also gives the vegetable consumption per head. Tomatoes stand out at 14.7 kg per head and onion consumption is also high at 6.9 kg but cabbages as a group approach this figure.

Vegetable consumption and expenditure by social class is given in Table 7. Expenditure on vegetables is high among the high income and low income groups, but the middle income group spends much less on vegetables. Purchases in kg on cauliflowers and cabbages are highest in the low income group, who also buy much the most lettuce, tomatoes and onions. The low income group also buy more cucumbers and carrots. Peppers are the only commodity which reverses the trend. Increasing amounts are spent on peppers with increases in income.

Table 7. Fresh Vegetable Consumption and Expenditure by Income Group - 1989

	Low income group	Middle income group	High income group
Purchases (kg)			
Cauliflower	2.21	1.01	1.27
White cabbage	1.36	0.84	0.91
Red cabbage	0.62	0.35	0.26
Lettuce	2.47	1.28	1.61
Asparagus	0.99	0.28	0.61
Tomatoes	5.98	3.53	4.82
Cucumbers	3.70	2.89	3.40
Peppers	1.31	1.50	2.12
Onions	5.45	2.94	3.35
Peas and beans	0.77	0.20	0.37
Carrots	3.83	2.25	3.46
Radish	0.86	0.38	0.59
Purchases (DM)	**104.70**	**57.87**	**91.29**

Source: ZMP Bilanz - Gemüse 1987-88

Modern and competitive marketing is a feature of the fruit and vegetable marketing in West Germany where the supermarket has a bigger role than in Britain. The devastation of city centres in the last War gave West Germany an advantage in re-siting her city markets. Again the dispersion of cities over West Germany has done much for market penetration in her lucrative market for fruit and vegetables.

The West German Market for Fruit

West German trade in fruit is given in Table 8. Data on the market for fruit is presented in the ZMP reports in three parts: the first part deals predominantly with temperate fruit, referred to as "Frischeobst" the second part with citrus fruit and the third part with "Southern fruit" or "Sudfruchte" substantially from overseas. Of the fruit trade figures in Table 8, fruit exports, as with vegetables, repre-

Table 8. West German Trade in Fruit - 1988, tonnes

	Imports	Exports
Table grapes	274579.0	9311.8
Wine grapes	2133.2	118.8
Water melons	77174.9	1015.2
Sweet melons	30922.8	430.7
Cider apples	22683.1	1374.9
Table apples	583997.0	29554.8
Perry pears	1576.8	17.7
Table pears	128704.5	5552.1
Quinces	1039.3	31.3
Apricots	27547.1	1895.9
Sweet cherries	16719.8	3122.9
Sour cherries	7064.3	1293.8
Peaches	116358.9	2241.1
Nectarines	110998.2	2248.0
Plums	23101.4	5008.4
Other stone fruit	126.0	34.3
Strawberries	100052.9	1428.4
Raspberries	7509.2	979.2
Loganberries	903.2	103.1
Blackberries	6187.8	182.5
Redcurrants	4042.0	235.2
Gooseberries	4105.3	60.1
Cranberries	1477.9	167.4
Bilberries	791.3	107.8
Bog bilberries	603.7	13.0
Other berried fruit	40.1	9.7
Kiwi fruit	53797.3	1108.3
Other fruit	6902.0	400.4
Fresh fruit	**1611150.0**	**66946.0**
Blood oranges	30495.1	433.0
Navel oranges	407185.9	13079.5
Other sweet oranges	11665.0	769.6
Bitter oranges	389.5	49.1
Clementines	104821.4	1754.3
Monreales and satsumas	40789.6	140.6
Mandarins and wilkings	3604.6	41.5
Tangerines	5022.0	190.5
Other citrus	7956.0	609.6
Lemons	101699.9	1653.4
Limes	329.1	11.5
Grapefruit	79432.2	13149.7
Citrus fruit	**793390.3**	**31882.3**
Bananas	677906.5	27212.9
Figs	1258.4	89.6
Pineapples	32048.2	588.7
Avocados	5230.9	113.4
Guavas and mangoes	4224.6	136.5
Pawpaws	1545.0	127.0
"Southern fruit"	**72213.6**	**28268.1**

Source: ZMP Bilanz - Obst, 1987-88

Table 9. West German Imports of Fruit (excluding citrus) by Country of Origin - Quantity and Value 1987

Country	Quantity tonnes	Quantity per cent	Value '000 DM
France	272010	14.4	324091
Belgium/Luxembourg	32692	1.7	54091
Netherlands	70343	3.7	94606
Italy	854005	45.3	1143439
United Kingdom	8227	0.4	3101
Denmark	1670	0.1	1669
Greece	110606	5.9	146472
Portugal	754	0.0	1350
Spain	136335	7.2	255278
EEC	**1486642**	**78.7**	**2024407**
Yugoslavia	11263	0.6	6137
Austria	1433	0.1	2463
Switzerland	-	-	-
Turkey	9195	0.5	8265
Sweden	100	0.0	541
West Europe	**21991**	**1.2**	**17406**
Bulgaria	2387	0.1	2504
Czechoslovakia	5744	0.3	3103
Poland	8272	0.5	11405
Rumania	13319	0.7	12042
Hungary	34027	1.8	18507
USSR	1447	0.1	2877
East Europe	**65196**	**3.5**	**50438**
Israel	4641	0.3	13458
South Africa	93051	5.0	140579
USA	7288	0.4	13746
Canada	54	0.0	224
Chile	74111	3.9	108988
Mexico	487	0.0	1955
Argentina	33791	1.8	44160
Brazil	1018	0.1	1998
New Zealand	70337	3.7	161464
Australia	5865	0.3	8955
Other overseas countries	3279	0.2	7476
Overseas countries	**293922**	**15.7**	**504003**
Unallocated quantity	**16004**	**0.9**	**7760**
Total	**1883755**	**100.0**	**2604014**

Source: ZMP Bilanz - Obst 1987-88

Figure 2. West German Imports of Fruit by Country of Origin

Imports by Quantity – 1987

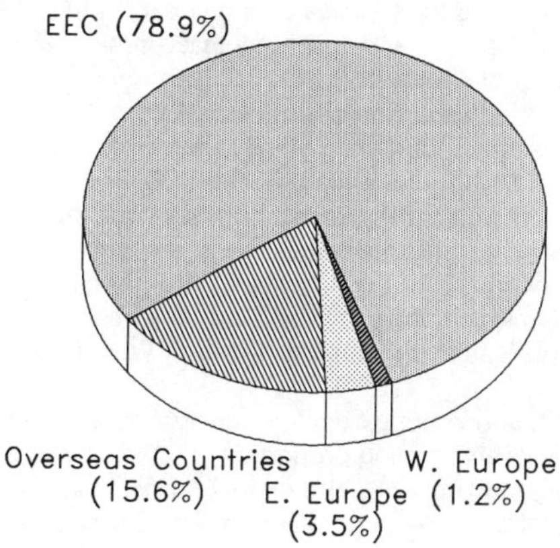

Imports by Value – 1987

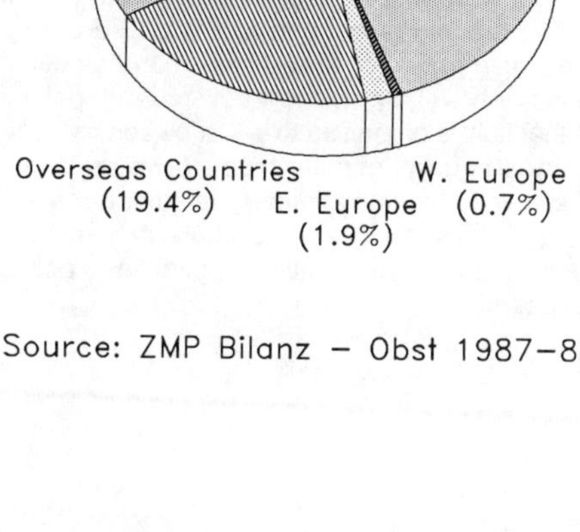

Source: ZMP Bilanz – Obst 1987–88

sentonly a small amount, four per cent of the imports. Fresh fruit (Frischobst) imports for 1988 were 1.6 million tonnes, citrus fruit nearly 0.8 million tonnes and "Southern fruit" over 0.07 million tonnes. Fresh fruit imported comprises a very wide range, the largest of which is dessert apples, 0.58 million tonnes, followed by pears, 0.12 million tonnes, peaches 1.1 million tonnes and nectarines 0.1 million tonnes. Citrus imports are mainly oranges, 0.61 million tonnes, and lemons, 0.1 million tonnes. Imports of bananas which vie with oranges as the largest import of all fruit produce run at 0.625 million tonnes. The wide range of produce imported and the large amount from various parts of the world indicate a profitable market outlet to the many countries concerned.

West German imports of fruit (excluding citrus fruit) by country are given in Table 9 and represented diagrammatically in Figure 2. Other partners in the EEC again supply most of these imports (78.9 per cent), while Italy alone supplies 45.3 per cent. Italy, in the West German fruit supply, contributes at the same level as Holland does for supplies of vegetables in West Germany. France is the next largest supplier of fruit with 14.4 per cent and Spain half that amount. Fruit imports from Portugal are negligible, but Greece has a stake of 5.9 per cent. Only 1.2 per cent of fruit imports are from 'West Europe', a rather curious description for the countries concerned.

Centrally planned economies which comprise chiefly East European countries supply 3.5 per cent of the fruit imports of which Hungary alone supplies half. Overseas countries have a notable presence in the imports, 15.7 per cent and by far the most of these fruit imports, 14.5 per cent, comes from prominent exporters in the southern hemisphere. These are South Africa (5.0 per cent), Chile (3.9 per cent), New Zealand (3.7 per cent) and the Argentine (11.8 per cent). A substantial proportion of these imports is apples and pears which arrive in West Germany fresh in their own season, and are preferred by consumers to the then less fresh supplies grown in West Germany or elsewhere in the northern hemisphere.

West German imports of fruit, and imports from the principal sources Italy and France are given in Table 10. This discloses that the overwhelming proportion of table grapes come from Italy and that Italy supplies more than twice as many cooking apples as France. Exports of the more profitable dessert apples from Italy are one third more than those from France, and Italy has seen competition with France in the apple trade rise in recent years. The West German - Italian connection in the trade in fruit and vegetables has been mentioned and also the importance of fruit in this connection. There is a tradition of exports from Italy to West Germany, using land routes through the Alpine passes to Munich as the centre of imports for this trade (The Munich Grossehandelmarkt). There is a similar tradition in the fruit and vegetable trade between Spain and Britain. With Spain in the Community, and Spain was disadvantaged in this trade before that, the stake of Spain in supplying fruit and vegetables to Britain will grow. Spain supplies 7.2 per cent of the imports of fruit excluding citrus and "Southern fruit" to West Germany, almost double that in 1981 when the Spanish supply was only 3.8 per cent of West German fruit imports. Imports of "Southern fruit" from Spain to West Germany went up from 30 per cent share in 1981 to a 37.7 share in 1987. The trend for Spain to get a bigger share of imports to West Germany will continue. Spain is likely to gain from exports to both Britain and West Germany at the expense of Italy whose climate and produce are very similar.

In Table 10 exports of dessert pears from Italy are only a little larger than those from France. There is an important production of summer pears in Italy for which she is well known. Italy has a large stake in the West German market for fruit and vegetables but if she is to hold her present advantageous place she will have to recognise the rising competition from Spain and develop her marketing to cope with it. The large Italian supply of peaches comes into its own, compared with France, by reasons of advantage in climate over that of France, and Italy is ahead of France in supply of other stone fruit. Italy again has a much bigger export of strawberries than France. Any one who has seen the double cover for strawberries in northern Italy - the row covered with polythene, then several rows covered together, can appreciate that Italy is organised to sell strawberries early in the season. This is followed up by heavy supplies of good quality of main crop strawberries. It is clear however, that West Germany draws supplies of strawberries from a number of other countries including Spain. The same goes for water melons and a number of other fruits where Italy dominates, but together with France, they do not supply the market completely. Italy supplies 45.3 per cent of fruit imports by value, and France supplies 14.4 per cent.

Table 10. West German Imports of Fruit from Principal Sources - 1987

	Total Imports	Imports from Italy	Imports from France
Table grapes	310332	202985	9693
Wine grapes	955	753	-
Cooking apples	98741	36202	16624
Dessert apples	672453	234393	161704
Cooking pears	2068	805	386
Dessert pears	141679	39924	37242
Quince	1767	-	425
Apricots	23761	5892	4542
Peaches	270364	221335	6648
Cherries	33907	10026	7087
Plums	36921	16871	9805
Other stone fruit	750	255	-
Strawberries	106710	41868	11211
Cranberries	1506	-	-
Bilberries	1155	-	22
Blackcurrants	4522	-	606
Redcurrants	2554	-	-
Raspberries	4166	133	72
Mossberries	382	-	-
Gooseberries	4718	28	-
Other berries	3157	-	-
Water melons	77068	29667	502
Sweet melons	34126	1893	2021
Kiwi fruit	44982	9832	3291
Other fruit	5013	1146	131
Total [a]	1883757	854008	272012
Per cent		45.3	14.4

[a] Fresh fruit excluding citrus and "Southern Fruit"

Source: ZMP Bilanz - Obst 1987-88

West German imports of "Southern fruit" by country of origin are featured in Table 11. Other member states of the EEC supply 42.5 per cent of this more exotic fruit, with Spain alone supplying 37.7 per cent. Mediterranean countries supply another 12.4 per cent with Israel in the lead and Morocco second. What ZMP refers to as the banana republics in Central and South America make a big contribution in the supply of "Southern fruit". The export is mainly bananas. Bananas are the market leader in West Germany. Other countries contribute 6.2 per cent of the supplies of "Southern fruit" of which Africa and the Ivory Coast are the most prominent. Imports of "Southern fruits" are featured in commodity detail and country of origin for Spain, Italy, Israel and Morocco in Table 12.

Fifty eight per cent of orange imports come from Spain, nine per cent from Italy and a similar amount from Morocco. Ninety per cent of mandarins and clementines also come from Spain as also 85 per cent of the lemons. There is a rise in demand for mandarins and clementines and similar 'easy peelers' known to the French as 'petites frites'. The smaller import of the more exotic "Southern fruit" comes from distant countries with a warmer climate than the countries given in the table. Overall Spain contributes 37.7 per cent of this import, Israel 6.2 per cent, Morocco 4.7 and Italy 3.5. So Spain enjoys a strong position in this trade which she will hold and probably strengthen. Italy, with the same

Table 11. West German Imports of "Southern Fruit", 1987

	Quantity tonnes	Quantity per cent	Value '000 DM
France	5651	0.3	6129
Italy	63228	3.5	57495
Greece	18986	1.0	15674
Spain	690948	37.7	579999
EEC Countries	**778813**	**42.5**	**659297**
Cyprus	21843	1.2	17988
Israel	113093	6.2	102299
Morocco	83611	4.6	71395
Canary Islands	331	0.0	764
Turkey	7798	0.4	8006
Tunisia	553	0.0	368
Mediterranean Countries	**227229**	**12.4**	**200820**
Costa Rica	152364	8.3	153991
Ecuador	139956	7.6	161139
Guatemala	9059	0.5	9468
Honduras	90548	4.9	104749
Columbia	116766	6.4	125184
Panama	191606	10.4	233057
Nicaragua	14982	0.8	14827
Banana Republics	**715281**	**38.9**	**802415**
Ivory Coast	32917	1.8	37441
Kenya	731	0.1	1699
Swaziland	3040	0.2	2953
Argentina	3993	0.2	4182
Brazil	5540	0.3	12369
South Africa	35609	1.9	35216
Uruguay	1086	0.1	1041
USA	18452	1.0	23158
Other overseas countries	11657	0.6	16005
Other Countries	**113025**	**6.2**	**134064**
Total	**1834348**	**100.0**	**1796596**

Source: ZMP Bilanz - Obst, 1987-88

Table 12. West German Imports of "Southern Fruit", 1987

	Total	From Spain	From Italy	From Israel	From Morocco
Bananas	698576	-	-	-	-
Oranges	574759	335194	53147	65738	57444
Mandarins and clementines	286190	256706	583	5871	26071
Lemons	115214	98283	9127	2097	-
Grapefruit	92556	244	158	1746	-
Other citrus fruit	11792	80	-	59	97
Pineapples	38878	-	-	-	-
Other "Southern fruit"	16383	442	213	7582	-
TOTAL	1834348	690949	63228	113093	83612
Per cent total imports		37.7	3.5	6.2	4.7

Source: ZMP Bilanz - Obst, 1987-88

climate and for many years a member of the EEC with a favourable market in West Germany has a high home consumption, but at the same time Italy does not appear to be vigorous in the export trade. Both Israel and Morocco enjoy concessions on the EEC tariff, more liberal that those towards Spain, before Spain joined the European Community in 1986.

In Table 13 citrus fruit is listed among "Southern fruit". Citrus fruit comprise 59 per cent by quantity and 51 per cent by value of the wider definition of imports of "Southern fruit" and gives the proportion of each type of fruit imported by quantity and value.

Table 13. West German Imports of "Southern Fruit" by Type, 1987

	Quantity per cent	Value per cent
Blood oranges	2.95	2.58
Navel oranges	27.60	20.62
Other sweet oranges	0.78	0.67
Other oranges	0.00	0.00
Clementines	11.49	12.96
Satsumas	3.79	2.58
Mandarins	0.19	0.22
Tangerines	0.13	0.15
Citrus hybrids	0.61	0.77
Lemons	6.28	5.73
Grapefruit	5.05	5.00
Other citrus fruit	0.04	0.10
Total citrus fruit	**58.91**	**51.38**
Bananas	38.08	43.59
Pineapples	2.12	2.43
Avocados	0.53	1.23
Guavas and mangoes	0.23	0.83
Fresh figs	0.06	0.23
Pawpaws	0.07	0.31
Other "Southern fruit"	**41.09**	**48.62**
Total "Southern fruit"	**100.00**	**100.00**

Source: ZMP Bilanz - Obst, 1987-88

Table 14 gives further details of the quantity of "Southern fruit" imported to West Germany. Pineapples and peaches lead the list with some 90 thousand tonnes each and apricots come next at 58 thousand tonnes. Apart from this produce all the other fruit named except for the citrus fruit are of temperate zone origin. The ZMP seems to have a rather elastic definition of "Southern fruit".

Table 14. West German Imports of "Southern Fruit" by Quantity, 1987

	Imports tonnes
Ginger	443.4
Grapefruit	249.7
Mandarins	35131.4
Other citrus fruit	2779.5
Pineapples	89793.0
Apples	14582.9
Pears	31412.9
Peaches	89232.5
Apricots	58304.5
Plums	4056.3
Sour cherries	28485.6
Sweet cherries	2135.6
Wine grapes	108.5
Strawberries	7782.0
Raspberries	2690.1
Other berries	3960.3
Other fruit	84391.4
Total	**454539.5**

Source: ZMP Bilanz - Obst, 1987-88

The West German fruit supply, self-sufficiency and consumption per head for all fresh fruit excluding citrus fruit is given in Table 15. (Supply of course is domestic production plus imports).

Table 15. West German Fruit Supply, Self-sufficiency and Consumption per Head, 1987-8 (excluding citrus fruit)

	'000 t	Self-sufficiency per cent	Consumption kg/head
Apples	1797.5	56.1	28.3
Pears	421.2	64.0	6.5
Sweet and sour cherries	248.9	88.6	3.9
Plums	339.1	90.0	5.3
Apricots	24.4	7.0	0.4
Peaches	289.3	7.7	4.4
Raspberries	240.1	93.7	3.8
Strawberries	157.9	32.2	2.3
Berried fruit	9.1	-	0.1
Table grapes	305.9	-	4.5
Bananas	685.2	-	10.8
Other fresh fruit	633.2	-	11.3
Total	**5232.0**	**40.2**	**81.3**

Source: ZMP Bilanz - Obst, 1987-8

This shows that in 1987-8 West Germany was 40 per cent self-sufficient for fruit, excluding citrus. In that year West Germany was not one hundred per cent self-sufficient for any single fruit, which differs from the situation for vegetables. She was near to self-sufficiency for the highly perishable cherries and raspberries and plums but she was little more than half self-sufficient for apples and some two thirds self-sufficient for pears. That West Germany is only 32 per cent self sufficient for the highly perishable fruit, strawberries, indicates superb cooling, packing and transport of strawberries from Italy and France and by the numerous other countries involved in the lucrative fruit and vegetable trade. As far as consumption is concerned, of the range of fruit imports given in Table 15, apples are seen to lead at 28 kg per head and bananas follow at 11 kg per head. (The figure for apples in Table 15 does include cooking apples). The consumption of unspecified fruit is also 11 kg per head. Pear consumption per head is 6.5 kg and that of plums 5.3 kg. Consumption per head for peaches is nearly eight kg and of table grapes 4.5 kg. In total West Germans eat a considerable amount of fruit, a total for those listed in Table 15 of 73 kg per head.

The supply and consumption of all citrus fruit is given in Table 16. Oranges head the list in consumption per head and the consumption of fruit as a whole is put at 15 kg per head. With citrus juice in terms of fresh fruit equivalent, consumption per head is raised to 33 kg so putting West Germany among the highest consumers of fruit in the world, if not the highest.

Table 16. West German Supply and Consumption of Citrus Fruit, 1987-88

	Supply '000 t	Consumption per head kg
Oranges	512	8.3
Clementines	249	4.1
Lemons	105	1.7
Grapefruit	79	1.3
All fresh citrus	**945**	**15.4**
Citrus juice	1096	17.9
Total	2041	33.3

Source: ZMP Bilanz - Obst, 1987-88

Table 17. West German Fresh Fruit Consumption and Expenditure by Income Group, 1987

	Low income group	Middle income group	High income group
Purchases (kg)			
Apples	14.94	11.49	14.13
Peaches	4.27	2.64	3.00
Cherries	0.53	0.24	0.42
Plums	2.04	0.47	0.68
Strawberries	3.07	1.27	1.93
Grapes	4.80	2.33	2.95
Oranges	15.37	7.84	9.25
Seville oranges	2.76	1.21	1.62
Other citrus fruit	3.65	1.10	1.71
Bananas	11.17	8.16	8.90
Purchases (DM)	**26.28**	**45.75**	**59.01**

Source: ZMP Bilanz - Obst, 1987-88

Consumption and expenditure by social class is given in Table 17. The outcome is seen to be rather different from vegetables where the lower and the higher income groups purchased more. For fruit the low income group buy much the least or 26 DM per head. The middle income group purchase 46 DM per head, while the high income group spends yet more, 59 DM per head. This is explained by the fact that fruit is rather more expensive than vegetables such that the low income group can afford much less. Their apparent preference for vegetables is determined by their low income.

West German Market Research

The distinctive seasonal market fluctuations for fruit and vegetables have led to great efforts, in storage, protected cultivation, imports and so on, to lengthen the season of various products in West Germany in order to establish a more balanced supply situation for this market. These measures as well as changes in eating habits have, in part, led to marked modifications in the formerly customary seasonal demand[3].

Dressler's paper pointed out the development in trends in seasonal demand and their determinant causes, over thirteen years by time series data. The results of the analysis indicate a distinct tendency towards evening out of differences in seasonal demand which must be attributable to the following. The increased purchasing power of the consumer in West Germany which favours relatively expensive off-season products, and there is an increase of imports and distribution of off-season products. The seasonal price differences have partially evened off and the customer has become accustomed to an all-the-year-round availability of the most important fruit and vegetables. Fresh produce is substituted to a certain extent by preserved products. The determinant causes in this connection are operated differently in each case according to the type of "household", or size of income. As to the future development of seasonal demand the following tendencies emerged. The seasonal demand behaviour of low income group is in contrast to the two other family income groups, bound by stronger determining factors of income and tradition, which make the reduction of differences in the seasonal demand more difficult. The seasonal demand of the middle income group and the high income group seems to be getting more similar. Since the rates of change in the last few years have been increasingly reduced no comparable changes are expected in the coming years.

According to von Alvensleben, in a growing economy, variables such as income and prices are losing relative importance as factors determining changes in the demand for food, while other factors such as consumer attitudes and preferences are gaining relative importance[4]. In a high income economy like West Germany, future changes in demand will mainly be caused by changes in consumer attitudes and preferences and not so much by changes in income and prices. Although the income elasticities of the demand for vegetables is rather low, for the more expensive items they are above average. The total demand for vegetables seems to have reached saturation point, however the assortment is changing and there is a clear tendency towards more diversity in demand. Bulk items such as cabbage, cauliflower and beans are losing relative importance, while onions and especially the 'other vegetables' category containing most of the Mediterranean vegetables (sweet pepper, egg plants and so on) are gaining popularity.

When wholesalers were asked to give their opinion on the external quality of tomatoes from different sources, the following scores were given (0 being the lowest score, 5 being the highest score).

Tomato Source		Tomato Source	
Holland	4.2	West Germany, open field	3.2
Belgium	4.1	Morocco	2.7
West Germany, glass	3.9	Italy	2.4
Canary Islands	3.8	Albania	2.2
Israel	3.7	Rumania	2.2
Spain, mainland	3.3	Bulgaria	2.1

[3] Long term changes in seasonal demand for Fruit and Vegetables Dressler H G pp 233-248 Acta Horticulturae No 97 1979

[4] Consumer and wholesale preferences for fresh vegetables, especially tomatoes in the Federal Republic of Germany. Alvensleben, Reimer von, pp 77-86 Optimal Yield Management, Edited by Dan Rymon Gower 1988.

So far as taste is concerned the Dutch tomato has an image far below the average of 11 classes of tomatoes from other sources to which it was compared. With respect to external quality the Dutch had the highest score. Again this was tested against eleven other sources of tomatoes. External quality is the most important point to the wholesaler in his decision to buy. Despite the poor taste image, the Dutch tomatoes received the highest ratings, when wholesalers were asked to give their opinion on the external quality of tomatoes from different sources. Many wholesalers do not believe that consumers notice taste differences between tomato origin. According to the experience of wholesalers, the external quality is the dominating buying criterion. It is believed that this is the reason that Dutch tomatoes are the most preferred tomatoes on the West German market.

As far as the handling aspect is concerned, wholesalers gave the different tomatoes the following score. (A high score is 1, a low score is 4.)

Tomato Source		Tomato Source	
Holland	1.0	Israel	2.3
Belgium	1.7	Italy	2.4
Canary Islands	1.8	Morocco	2.6
Spain, mainland	1.8	Rumania	2.9
West Germany	2.1	Bulgaria	3.2

Consumer preferences and attitudes vary greatly. An average or typical consumer does not exist. For this reason, it is useful to design marketing strategies taking account of different market segments. To maximise consumer satisfaction, the retail trade should offer a range of products. Many consumers have preferences for home-grown and open-field produce as against, for instance, produce grown under glass. For this reason it is suggested that labelling to inform the consumer of the country of origin and the type of production and the name of the country exporting the produce. Wholesaler preferences do not always coincide with consumer preferences. In particular, the consumers' image of Dutch tomatoes is not as good as many wholesalers believe. For wholesalers the external quality and handling aspects seem to be much more important buying criteria than taste.

The paper by Vierheilig and von Alvensleben presents some results of a survey carried out by the Institute for Horticultural Economics at Hanover University in the summer of 1986 in Hanover.[5] Almost 300 women were chosen at random and interviewed about their opinions and preferences regarding salads. In the summer, 80 per cent of the consumers eat green salads at least once a week, 19 per cent consume green salads nearly every day. Older people consume more than younger people and in households with a high income much more green salad is consumed than in low income households. Nearly every person knows lettuce. But only up to ten per cent of the consumers have eaten new varieties of the green salad range (oak leaf lettuce, red varieties of lettuce, frisee and so on). Young people like tomato salad, cucumber salad, and mixed salad best of all, while older consumers prefer cucumber salad, lettuce and green bean salad. The image of lettuce seems to be old-fashioned. It can be concluded that lettuce has already passed the peak of its life cycle and will be substituted more and more by iceberg lettuce and other types of salads which are new on the West German market.

---

5 Vierheilig, B.M. and Alvensleben, Reimer von, pp 351-354, Acta Horticulturae No 223, 1988.

CHAPTER 5
THE UNITED KINGDOM MARKET FOR FRUIT AND VEGETABLES

The United Kingdom Fruit and Vegetable Economy

Estimates for 1988 place the field vegetable area of the United Kingdom as 178 thousand ha and protected crops (glass and polythene) cover some three and a half thousand ha.[1] The area of orchard fruit is 34 thousand ha, and of soft fruit, 15 thousand ha. Of orchard fruit, 14 thousand ha is in dessert apples, 8.5 in culinary apples and nearly four thousand ha is in pears. From 1979 to 1988 the area of dessert apples has shrunk by a quarter. This comes about from the successful competition of the French in the UK market for dessert apples, a trend which has been apparent since the early 1970s and onwards. Britain's entry to the European Community in 1973 only hastened the devastating effects of French imports. The climate in France gives her dessert apple production a yield more than twice the level of Cox's Orange Pippin in England, and the French variety, Golden Delicious, has always been in strong demand, a variety eminently suitable to supermarket needs.[2] Over the same period culinary apples suffered an even greater contraction than dessert apples. This trend of contraction in orchard fruit in England must continue but what remains is the more viable part of the English apple and pear sector. The smaller industry continues to compete with the French though there is every sign that the advantage of the French will continue. Promotion does not offset high production costs. Again the area under soft fruit has shrunk by almost twenty per cent since 1979. This includes strawberries which alone comprise half the soft fruit area.

Annual production of dessert apples is now from 130 to 150 thousand tonnes and that of culinary apples from 100 to 120 thousand tonnes. Pear production lies around 40 thousand tonnes per year while that of soft fruit is some 90 thousand tonnes. The value of the output of fruit in 1988 was £211 million of which a quarter was dessert apples, 16 per cent culinary apples and six per cent pears. The value of output of soft fruit was £99 million, nearly half the total fruit output. Britain grows the north temperate range of field vegetables, which have also declined in area by eight per cent since 1979. Twenty per cent of the area is in roots and onions, thirty per cent in brassicas and towards thirty per cent are leguminous vegetables, most of which are peas for processing. The remainder of non-specified vegetables account for twenty per cent of the vegetable area. The total value of field vegetables was £531 million for 1988. Roots and onions account for 28 per cent of all vegetables by value and the brassicas for 36 per cent. Legumes account for 11 per cent and "other vegetables" for upwards of 25 per cent.

---

[1] Basic Horticultural Statistics for the United Kingdom HMSO 1979-1988

[2] All the orchard fruit is in England

Fruit in Britain

The area of fruit grown in Britain and the production by volume are given in Table 1.

Table 1. UK - Fruit Production 1987, Areas and Quantities

Items as listed in Eurostat

	'000 ha	'000 tonnes
Apples	27.0	312.0
Pears	4.0	66.4
Apples and pears	**31.0**	**378.4**
Cherries	0.9	3.6
Plums	2.7	35.6
Stone fruit	**3.6**	**39.2**
Black and red currants	3.1	17.9
Strawberries	6.4	53.8
Raspberries	3.8	24.0
Gooseberries	0.6	6.0
Other soft fruit	0.6	4.2
Soft fruit	**14.5**	**105.9**
Total of fruit specified	**49.1**	**523.5**

Source: Eurostat 1989

Apples and pears comprise 31 thousand ha out of a total fruit area of 49 thousand. This gives rise to a production of 523 thousand tonnes. Pears are in the minority compared with apples, but they are grown on farms with apple orchards, where they play a part in contributing to the viability of the orchard business on the farms concerned. A large proportion of the English apple and pear orchard is in Kent, some eighty per cent. Here there are good fruit soils and the most favourable climate for apples and pears in the UK. Most of the apple and pear enterprises are on farms with diverse farm enterprises which gives the Kentish apple and pear sector a business advantage over the specialised orchard farms in East Anglia. Also yields of apples and pears in East Anglia, at the northern limit of apple and pear production in England, are lower than those in Kent, which are in turn lower than those in France, which benefit from longer hours of sunshine with the more southern location.

Cherry orchards in England have declined in area continually since 1945 despite their most favoured location in Kent. Their decline is entirely due to the greater competition from the Mediterranean countries, Spain and Italy with some imports from France. In the stone fruit sector, the production and area of plums is three times that of cherries. Plums have the unfortunate characteristic of having a price variance from one to eight, the largest price swing of all produce in the UK. This is a consequence of considerably large variations in annual yields.

Soft fruit not only has four times the prominence of stone fruit in area, and almost three times the production of stone fruit by volume, but soft fruit has been much more profitable over the years than any other type of fruit grown in the UK. This situation is substantially due to the fact that soft fruit is highly perishable, which gives UK growers an advantage in their own harvest season. The climate in the UK favours soft fruit production and as soft fruit production is also land intensive, so the area of land can be small, it is not difficult to find favourable sites. Strawberries, which may be found growing commercially in practically all parts of Britain, are mainly grown in central and southern England.[3] Some eighty per cent of the raspberries are grown on a large scale on mixed cropping farms in the East of Scotland where the climate is eminently suitable. Because of this, the UK grows more raspberries than any of her counterparts in the EEC. Again, the market of fruit for processing supports the strawberry economy even though jam making in Britain has declined. For a number of years the sector of strawberries for processing has been under pressure from cheap imports of strawberry pulp from Poland. Despite UK Farmers Union demands that the UK government should protect the sub-sector nothing was achieved in this direction as Brussels did not support the idea. However a study of the Economics of Strawberries for Processing commissioned by the UK government, found that since twenty years ago the yield of strawberries for processing has doubled while the area has halved.[4] The study, relating to the 1988 crop, showed that the sub-sector is actually quite profitable, the smaller, family growers gaining because they incur little paid labour, if any, while the larger scale growers gain from economy of scale.

While the outlook for strawberries is good, the outlook for the remaining other soft fruit is also good, because so many producers have gone out of production. For example, supplies of red and white currants are so low that consumers have a job to find them in the markets. Black currants are grown on a number of arable farms in East Anglia under contract with Ribena and such firms. Ribena produces a high vitamin concentrate. Producers for processing have contracts renewed from season to season by the manufacturers on a cost plus basis.

"Pick-your-own", is probably more important in the UK than in any other country in the EEC. This is because of the high degree to which the production net-work throughout the country may be made to fit the distribution of the population. Many of the larger units of production, of course, sell only a part of their output in this way. All soft fruit and apples and pears feature the pick-your-own system. The system is popular because price is set at a level advantageous to both producer and consumer and the consumer gets the produce in the fresh state. The producer avoids altogether the high cost of picking, and a great number of producers over the country have set up their farms to cash in on the pick-your-own system. Because of perishability, and lower, but profitable prices in the home season, when imports cannot compete, and also, because of pick-your-own and good yields, as has been stated, the soft fruit sector in Britain is the most profitable fruit sector. For a number of reasons, but in great measure because of effective competition from France, the apple sector in Britain is the least profitable.[5] Competitive pressure from France in the dessert apple sector, already great before 1973 when Britain acceded to the EEC, which established free trade between Britain and France, has become greater from this time.

Vegetables in Britain

The area of vegetables and the production by volume are given in Table 2. Vegetables in the UK are three times the area of the fruit sector and more than six times the production by volume. Field vegetables, many of which are grown in the UK are large scale, are viable as a sector in Britain and this sector will continue to be viable in future. This has brought about a situation in the vegetable sector of self-sufficiency in supply of field vegetables of at least eighty per cent. Low cost production stems from a large scale operation and a climate ideally suited to providing good yields of quality vegetables. Then there is the fact that vegetables are bulky and of relatively low value compared with fruit, such that the cost of transport of vegetables from other member states or elsewhere often means that the cost of transport gives little return to exporters. For example it costs £30 per tonne to transport potatoes from Holland to the UK. More valuable commodities like lettuce in the winter, are of

---

[3] English Strawberries. An Economic Study of the Production and Marketing of Strawberries. Hinton, Lynn. Occasional papers No 23. Department of Land Economy, University of Cambridge, February 1981.

[4] The Economies of Strawberries for Processing, Hinton, Lynn, Department of Land Economy, University of Cambridge, December 1988.

[5] Economic Results from Horticulture, Hinton, Lynn and Housden, Walter. Agricultural Economics Unit, Department of Land Economy, University of Cambridge.

course excepted. Adding to the viability of vegetables in Britain, is that the farms growing vegetables usually have other important crop enterprises as well. The range of diversification of production of other arable crops on farms with vegetables is much to the producers' advantage.

Table 2. UK - Vegetable Production 1987, Areas and Quantities

Items as listed in Eurostat	'000 ha	'000 tonnes
Cauliflower	20.4	390.5
Brussels sprouts	10.8	160.8
Cabbage	22.4	692.3
Other brassicae	-	0.1
All brassicas	**53.6**	**1243.7**
Celeriac and celery	1.1	64.0
Leeks	3.1	71.5
Lettuce	7.2	224.3
Asparagus	0.6	1.1
Other vegetables (leafed or stalked)	0.9	31.5
Vegetables leafy or stalked	**12.9**	**392.0**
Tomatoes	0.6	129.1
Cucumbers	0.2	79.0
Peppers	0.1	2.5
Cultivated for fruit	**0.9**	**31.5**
Turnips	4.8	166.0
Carrots	14.7	567.9
Onions	9.5	317.1
Beetroot	2.6	100.8
Other roots and tubers	3.0	62.2
Roots and tubers	**34.6**	**1214.0**
Peas	41.6	187.7
French beans and runner beans	7.2	50.8
Other pulses	2.1	10.7
Pulses	**50.9**	**249.2**
Cultivated mushrooms	**0.5**	**103.9**
Other fresh vegetables	**12.2**	**154.2**
Total of vegetables specified	**165.6**	**3568.0**

Source: Eurostat 1989

Brassicas alone are one third of the total area and production of vegetables in the UK. Among the leafed or stalked vegetables, which cover 13 thousand ha, lettuce is the most important crop, followed by leeks. Part of the lettuce crop of course is grown under glass when the producer might grow as much as seven crops in a year. The rest is grown in the open to meet the summer demand for salads and there may be a succession of lettuce crops. Carrots stand out with the largest area of roots and tubers and the largest value of production. Most of the peas are grown on contract for freezing but the area of peas grown for marketing fresh is small and declining. The total area of vegetable crops under glass in Britain is 3403 ha, of which lettuce is half. Glasshouse tomatoes, cucumbers and peppers comprise almost one thousand ha of vegetables cultivated for their fruit, and tomatoes, because of effective competition from the Dutch, now account for only one fifth of the area down to glasshouse vegetables.

Mushrooms in Britain are a growth sector somewhat similar in expansion to the broiler chicken industry. Mushroom production per square foot has almost doubled in the last twenty years. For some years now the value of mushroom output in Britain, as mushrooms increased in production, have been greater than the value of the output of tomatoes, as this crop has declined.[6 7] Mushroom production in Britain is highly competitive with some two hundred or so producers among which a handful of very large producers account for half of the total production of the mushroom industry. The other producers are mainly family units. What is quite remarkable is that the mushroom sector is profitable on mushroom units of both types. The very large unit with economies of scale and the family business which gains from the results of close supervision and profit-led family management. Such units also stand to get a better price by selling to local wholesalers in their own locality.

UK Imports of Vegetables

UK imports of vegetables are given in detail in Table 3 and they are illustrated in Figure 1. Those from our partners in the EEC amount to 638 thousand tonnes, and from all sources they are 873 thousand tonnes. This means that virtually three quarters come from our partners in the EEC. Thirty one per cent of all UK vegetable imports come from Holland and marginally less from Spain. Next in prominence, but a long way behind Spain and Holland, is France with seven per cent of UK vegetable imports, and Italy with 3.9 per cent. Much of the smaller import from West Germany represents entrepôt trade.

Of the total imports of tomatoes of 256 thousand tonnes, sixty per cent or 152 thousand tonnes come from the EEC (110 thousand tonnes from Holland) and 103 thousand tonnes or 40 per cent come from the Canary Islands and Spain over the winter months. Britain is heavily dependent on Holland to supplement her domestic supply of salad vegetables. Not only does Holland supply over 100 thousand tonnes of tomatoes, but she also supplies 23 thousand tonnes of cucumbers and 13 thousand tonnes of sweet peppers. Holland supplies over one quarter of the UK's large import of onions. It is a tribute to Dutch expertise in growing onions that Holland has such a large stake in the UK market in competition with our own large onion production.

Spain supplies 29 per cent of the vegetable imports to Britain. Spain is an important supplier of tomatoes, some way behind Holland and the Canary Islands. Spain supplies 39 thousand tonnes of tomatoes over eight thousand tonnes of cucumbers, nearly 13 thousand tonnes of sweet peppers together with a diverse and large amount of field vegetables. Half of Britain's imports of onions come from Spain. In the Spanish export season climatic differences over Spain's land mass play a part. Onions are harvested in the north in June and the harvesting season progresses with supplies from the middle of Spain and from regions further south, such that Spain can offer fresh onions for some four or five months of the year.[8] While Spain and Holland together supply over three quarters of onion imports to Britain the remainder come from many parts of the world which have a climatic advantage in ripening them. In 1987 the world export supply of onions to Britain was some 240 thousand tonnes, 192 thousand from the EEC. The balance of imports were made up by Poland, 19,159 tonnes, Czechoslovakia 595 tonnes, Hungary 5,345 tonnes, New Zealand 2,690 tonnes, Australia 4,998 tonnes, Chile 9,946 tonnes, South Africa 397 tonnes, Egypt 194 tonnes and Argentina 86

---

[6] Mushrooms in Britain and Ireland, Hinton Lynn, Department of Land Economy, University of Cambridge, 1982.

[7] The Economics of Mushrooms 1985, Hinton Lynn, Open Learning Teaching Module. Agricultural Training Board 1986.

[8] EEC Enlargement in the Fruit and Vegetable sector with Special Reference to Spain and the United Kingdom. Occasional Papers No 31 Cambridge University, Hinton Lynn, November 1983.

Table 3. UK Imports of Vegetables by Country of Origin, tonnes 1987

	TOTAL	FR	BL	NE	IT	IR	SP	WG	GR	EEC	PER CENT	INTRA EEC	OTHER COUNTRIES	PER CENT
Gherkins	169			28			120		21	169	100.0	172		
Cabbage lettuce	52994	2967	353	11756	2759		33952	10		51797	97.7	52906	1197	2.3
Witloof chicory	656	24	195	166	271					656	100.0	659		
Tomatoes	256144	432	923	110339	57	978	39035	10		151774	59.3	151941	104297	40.7
Cucumbers	53520	133	1	23120			8355	4	540	32153	60.1	32189	21367	39.9
Fennel	982			122	794		66			982	100.0	1075		
Sweet peppers	36083	187	30	12866	698		12817	15		26613	73.8	26620	9470	26.2
Other salad vegetables	2549	1145	228	541	522		113			2549	100.0	2584		
Peas	1701	86		21	219		150			476	28.0	530	1225	72.0
Beans	6349	93	7	268	118		1288			1774	27.9	1932	4575	72.1
Other legumes	189	50		39	50		50			189	100.0	189		
Horse radish	80											145	80	100.0
Cauliflowers	40907	21914		338	9285		9370			40907	100.0	40935		
Brussel sprouts	1852		40	1704			108			1852	100.0	1899		
Cabbages red and whit	32075	1818		25127				5130		32075	100.0	32444		
Other cabbages	8837	1263		846	1851		1691			5651	63.9	5667	3186	36.1
Garlic	4156	630		46	602		2121			3399	81.8	3423	757	18.2
Leeks	1345	162	16	908	137			122		1345	100.0	1476		
Asparagus	1352	28		29			315			372	27.5	376	980	72.5
Spinach	504			1	503					504	100.0	627		
Onions	235275	7644	119	61264	1865		116837	4047		191776	81.5	192786	43507	18.5
Mushrooms	23754	161	3253	9615	4	10718		4		23755	100.0	23659		
Courgettes	11557	2678		353	1681		6845			11557	100.0	11583		
Aubergines	4737	2747			103		97			2947	62.2	2984	1790	37.8
Celeriac	213			213						213	100.0	638		
Carrots and turnips	42674	14745	152	5875	9920		5679	19		36390	85.3	36457	6274	14.7
Other vegetables	36676	423	68	1179	1769		13208	2		16649	45.4	16973	20027	54.6
TOTAL	857256	59330	5385	266764	33208	11696	252217	9363	561	638524	74.5	642869	218732	25.5
Percent specified countries of total		6.9	0.6	31.1	3.9	1.4	29.4	1.1	0.1	74.5			25.5	

Source: Eurostat External Trade 1989 - Imports

Figure 1. UK Imports of Vegetables by Country of Origin

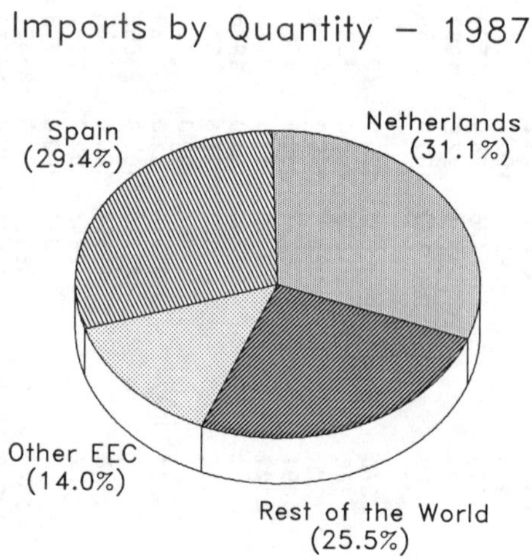

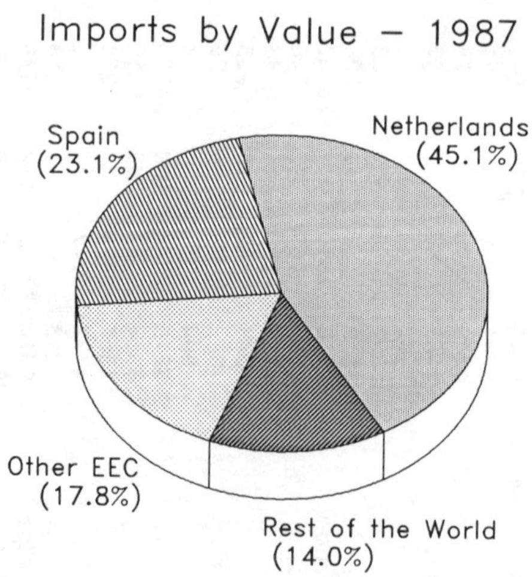

Source: Eurostat External Trade 1987 – Imports

tonnes. Imported onions gain a premium on the British market which justifies the freight cost for many distant exporters. Though Britain has become rather more self-sufficient in her supply of onions over the last twenty years, the increased self-sufficiency has levelled off in recent years at around fifty per cent. Greater self-sufficiency from home production, though less spectacular than export success, truly gives more scope to the vegetable industry in Britain, than marginal exports of a few commodities. The present balance of imports and home production in Britain gives a measure of competitiveness between British producers and those producers exporting successfully in the British market.

UK Imports of Fruit, excluding Citrus Fruit

Fruit imports to Britain are shown in three separate tables, one for fruit other than citrus fruit, one for citrus fruit and one for bananas. To deal with the majority of fruit imports first, though much of the fruit is temperate in origin, the items of produce given in Table 4 also includes exotics such as dates, figs and pineapples. Britain depends on her partners in the EEC to the extent of 68.3 per cent of the fruit shown in Table 4. Import shares by volume and value are illustrated in Figure 2. Britain is over 90 per cent dependent on the EEC for apricots, peaches and nectarines and for cherries.

Most prominent among the imports are the temperate tree fruits, apples, pears and plums. Britain is over sixty five per cent dependent on her partners in the EEC for imports of apples and some seventy per cent for pears and the more perishable plums. France alone accounts for more than half of the apple imports to Britain and significant quantities also come from Belgium-Luxembourg, the Netherlands, Italy and Spain. Of extra-EEC exporters of apples to Britain South Africa is the most prominent, sending 78 thousand tonnes, while Chile sends 15 thousand tonnes and New Zealand 27 thousand tonnes. Southern hemisphere apples are important in the British market because they arrive fresh in the off-season. Seventy-two per cent of the smaller import of pears come from partners in the EEC and almost thirty per cent of these come from France. Other large sender of pears are the Netherlands with 16 thousand tonnes, and Belgium-Luxembourg with seven thousand tonnes, followed by Italy and Spain.

Imports of table grapes are ten per cent of all the imports shown in Table 4. Sixty-three per cent come from the EEC, of which Italy and Spain are the most important suppliers, each sending some twenty thousand tonnes of total imports of 97 thousand tonnes. Next in importance in the imported supply is South Africa, followed by Greece, Cyprus and the USA. The consignment from Holland is entrepôt trade, which like the exports of oranges from Holland does not of course represent Dutch production. The import of dates from France also represents the entrepét trade of her former dependencies. The biggest suppliers of dates to Britain are Iran and China, and these two sources together account for over half the imports of dates. European entrepôt trade also features in the supply of figs and pineapples. Of the 21 thousand tonnes of pineapples 13 thousand tonnes comes from the Ivory Coast. Next comes Ghana with over two thousand tonnes, then the Dominican Republic, followed by other countries in Africa and South America. Most of the small import of papaws comes from Brazil.

UK Imports of Citrus Fruit

Citrus fruit imports to Britain are shown in Table 5. Import shares by volume and value are illustrated in Figure 3. Sixty per cent of the imports of citrus fruit comes from the EEC. Spain is by far the largest contributor and Spain alone accounts for 55 per cent of the supply. Other imports from the EEC, apart from small imports from Italy and France are entrepôt trade. Spain supplies virtually all the monreales and satsumas and a quarter of the large imports of the common orange, which alone comprises more than one third of British imports of citrus fruit. A quarter of the lemons imported come from Spain as do half the oranges. Non-EEC countries export 68 per cent of the oranges and over fifty per cent of the sweet oranges and other than sweet oranges, they also supply 60 per cent of the sanguines and almost thirty per cent of the tangerines. EEC trade in lemons amounts to 62 per cent of the total lemon imports of 51 thousand tonnes, Cyprus is prominent with 12 thousand tonnes and South Africa with five thousand tonnes. Ninety four per cent of the grapefruit is extra-EEC trade. The eighty thousand tonne import of grapefruit comes from many countries. Cyprus is the most prominent exporter with 33 thousand tonnes followed by Israel with fifteen thousand tonnes and South Africa with thirteen thousand tonnes.

Table 4. UK Imports of Fruit, other than Citrus Fruit and Bananas, by Country of Origin, tonnes 1987

	TOTAL	FR	BL	NE	IT	SP	WG	GR	PO	EEC	PER CENT	INTRA EEC	OTHER COUNTRIES	PER CENT
Dates	9482	1190	5	17						1212	12.8	1348	8270	87.2
Figs	302	26			1					27	8.9	56	275	91.1
Pineapples	20681	1006	695	403			14			2118	10.2	2149	18563	89.8
Kiwi fruit	5381	157	86	326	537					1106	20.6	1220	4275	79.4
Avocados	18510	191	45			175				411	2.2	418	18099	97.8
Mangoes, guavas and mangosteens	8125	40	7	137						184	2.3	244	7941	97.7
Apricots	3186	1152			351	821	5	596		2925	91.8	2976	261	8.2
Peaches and nectarines	94739	10879	38	888	74613	5387	17	2351		94173	99.4	94246	566	0.6
Cherries	7235	4908	223	240	1698	9	5	48		7131	98.6	7309	104	1.4
Other stone fruit	0											298		
Apples	440599	247999	15035	9799	8647	4097	2116	18	18	287729	65.3	288806	152870	34.7
Pears	75223	21729	7252	16028	5336	3910	2			54257	72.1	54388	20966	27.9
Plums	23974	4179		1476	2626	7964				16245	67.8	16304	7729	32.2
Quinces	12												12	100.0
Table grapes	97432	613	760	7663	21440	21622	139	8764		61001	62.6	61090	36431	37.4
Pawpaws	846		1	16						17	2.0	20	829	98.0
Strawberries	14863	1654	578	904	744	9753			487	14120	95.0	14250	743	5.0
Raspberries and redcurrants	143		5	1						6	4.2	23	137	95.8
Cowberries,foxberries	0											2		
Blackcurrants	0											167		
Cranberries	103			1						1	1.2	23	102	99.0
Other fresh berries	52	3		11						14	26.9	79	38	73.1
Water melon	17018	117		263	2058	12190		2164		16792	98.7	16807	226	1.3
Other fresh melon	76250	3035	116	2121	690	59447				65409	85.8	65819	10841	14.2
Other fresh fruit	6097	65	1	74	80	3240				3460	56.7	3460	2637	43.3
TOTAL	920253	298943	24847	40368	118821	128615	2298	13941	505	628338	68.3	631502	291915	31.7
Percent specified countries of total		32.5	2.7	4.4	12.9	14.0	0.2	1.5	0.1	68.3			31.7	

Source: Eurostat External Trade 1989 - Imports

Figure 2. UK Imports of Fruit, other than Citrus Fruit and Bananas, by Country of Origin

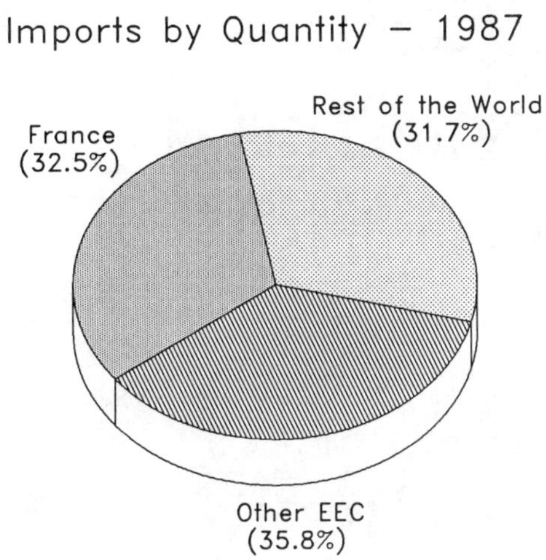

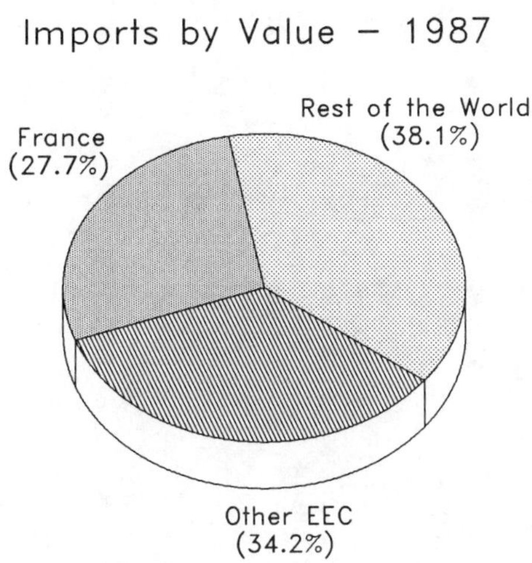

Source: Eurostat External Trade 1987 – Imports

Table 5. UK Imports of Citrus Fruit by Country of Origin, tonnes 1987

	TOTAL	FR	BL	NE	IT	SP	WG	GR	EEC	PER CENT	INTRA EEC	OTHER COUNTRIES	PER CENT
Oranges	303175	1892	1750	10900	943	76032	1098	3818	96433	31.8	97786	206742	68.2
Other sweet oranges	11340			2931	1410	781			5122	45.2	7858	6218	54.8
Other than sweet	7510			110		3446			3556	47.4	4290	3954	52.6
Sanquines	2348				620	323			943	40.2	1125	1405	59.8
Clementines	260347	2709	451	562		248811	20		252553	97.0	29795	7794	3.0
Monreales and Satsumas	91885	184				91588			91772	99.9	92369	113	0.1
Mandarins and Wilkins	2475			714	633	1111			2458	99.3	2580	17	0.7
Tangerines	4388			2670		438			3108	70.8	3146	1280	29.2
Other citrus hybrids	4276			475		1103			1578	36.9	2650	2698	63.1
Lemons	50670	1167	134	696	387	29220			31604	62.4	32053	19066	37.6
Grapefruit	80441	784	269	1090		1107	1516		4766	5.9	5150	75675	94.1
Other citrus fruit	667			17					17	2.5	45	650	97.5
TOTAL	819522	6736	2604	20165	3993	453960	2634	3818	493910		278847	325612	
Percent specified countries of total		0.8	0.3	2.5	0.5	55.4	0.3	0.5	60.3			39.7	

Source: Eurostat External Trade 1989 - Imports

Figure 3. UK Imports of Citrus Fruit by Country of Origin

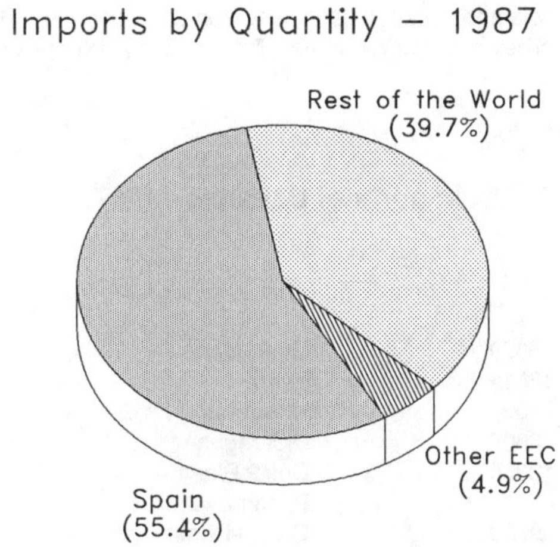

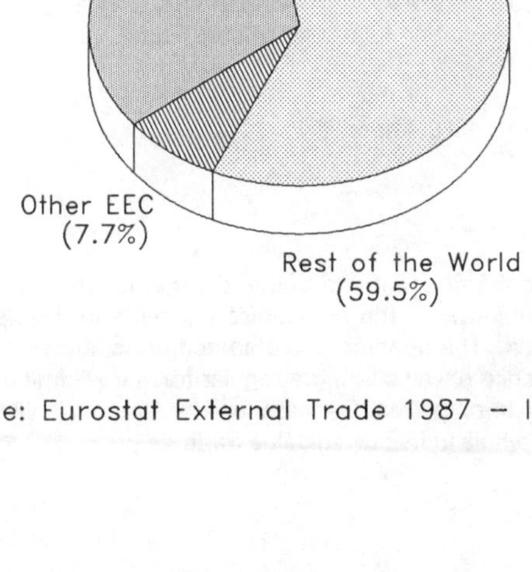

Source: Eurostat External Trade 1987 – Imports

Bananas

World trade in bananas is of particular significance to the economy of a number of developing countries in the tropical belts of the world, particularly those in Central and South America. The UK imports of bananas are given in Table 6. The eleven per cent from European sources represents entrepôt trade. The imports from Africa are so small (0.09 per cent of imports) as to be of little consequence but they do indicate the potential for African exports of bananas. Practically all the bananas exported to Britain come from fifteen countries in Central or Southern America or from Islands in the Caribbean. St Lucia and Dominica together account for nearly forty per cent of banana exports to Britain, while Jamaica, Colombia and Surinam each send some ten per cent. Bananas are the largest single fruit import to the UK. They account for some 17 per cent by volume and 16 per cent by value of all fruit imports to the UK.

Table 6. UK Imports of Bananas - 1987, tonnes

		per cent			per cent
France	5520		Guatemala	706	0.20
Belgium/Luxembourg	13018		Belize	19445	5.00
Netherlands	5027		Honduras	4347	1.00
West Germany	2805		Nicaragua	153	0.04
Ireland	13368		Costa Rica	2145	0.60
			Panama	4862	1.00
Intra EEC Trade	**39738**	**11.00**	**Central American Trade**	**31658**	**9.00**
Canary Islands	8		Colombia	33590	9.00
Ivory Coast	223		Surinam	33869	9.00
Cameroon	59		Ecuador	903	0.25
Somalia	38		**South American Trade**	**68362**	**19.00**
African Trade	**328**	**0.09**	Dominica	60859	17.00
			Martinique	1225	0.34
			Jamaica	34379	10.00
			St. Lucia	78381	22.00
			St. Vincent	35829	10.00
			Grenada	8383	2.00
WORLD TRADE	359142		**Caribbean Trade**	**219056**	**61.00**
EXTRA EEC TRADE	319680		**American and Caribbean Trade**	**319076**	**89.00**

Source: Eurostat External Trade, 1989, Imports

Bananas have an important function in the fruit and vegetable trade. Boat arrivals and the quantity in transit of bananas are known. The companies concerned have banana ripening rooms distributed throughout the Britain. The ripening and dispatch of bananas is regulated. This gives companies with bananas the chance of establishing a regular trade in all fruit and vegetables "on the backs of bananas" so to speak. In both West Germany and Britain this is very important feature of distribution and marketing in the whole fruit and vegetable trade.

The Market for Fruit and Vegetables in the UK

Data from the National Food Survey for 1987 puts expenditure on fruit and vegetables in the UK at 21.7 per cent of expenditure on all food, second only to meat and meat products and a quarter more than expenditure on cereals. This represents a substantial change over the position in 1965 when the proportion spent on fruit and vegetables amounted to only 17 per cent of total food expenditure.[9]

The Market for Fruit

There are many similarities in the trends in the consumption patterns for fruit as there are for vegetables. The range of fruit available to the British consumer has increased greatly over recent years. Nevertheless, traditional fruits such as apples, oranges and bananas continue to make up the bulk of the purchases. Fresh fruit has increasingly replaced canned fruit and the consumption of fruit juice is increasing.

Fresh fruit expenditure has nearly trebled since 1968 and it now accounts for some 8.6 per cent of total household food expenditure. The consumption data for apples, oranges and bananas shows no clear trend although levels are somewhat lower in 1987 than in 1965. At the same time, purchases of other citrus fruit, stone fruit and miscellaneous fresh fruit have increased markedly. This follows from the increasing popularity of the less traditional and more exotic varieties of fruit such as kiwi fruit, passion fruit and clementines and nectarines.

The Market for Vegetables

Over the last 20 years, the range of vegetables purchased has greatly increased. The availability of certain types of vegetables such as peppers, yams and garlic sought after by a more cosmopolitan population has increased. At the same time, traditional vegetables, such as potatoes, cabbage and carrots, remain important components of the consumption of vegetables.

Table 7. The Consumption of Vegetables

	1965	1975	1987	1965	1975	1987
	grammes/person/week			kg/person/year		
Fresh potatoes	1656	1365	1172	86	71	61
Fresh green vegetables	446	360	310	23	19	16
Other fresh vegetables	446	429	518	23	22	27
Frozen vegetables and vegetables products	27	101	209	1	5	11
Other processed vegetables and vegetable products	306	357	392	16	19	20
Total	**2881**	**2612**	**2606**	**150**	**136**	**136**

Source Household Food Consumption and Expenditure 1987 HMSO 1989

[9] Household Food Consumption and Expenditure 1987 HMSO 1989

Fresh Potatoes

Although potatoes remain an important vegetable in the diet in Britain, their relative importance, as one might expect has declined steadily over the last 20 years as shown in Table 7. Household consumption averaged 1656 grammes per person per week in 1965 and represented 57 per cent of total vegetable consumption in that year. By 1987 the average per capita consumption had fallen to some 1172 grammes accounting for 45 per cent of all vegetables consumed. Some of this decline was offset by the growth in sales of potato products which now account for some 342 grammes in fresh equivalent terms.

The price paid for potatoes has remained relatively stable in real terms, and for most of the period in which there was an underlying downward trend in consumption. A second successive very dry summer in 1976 resulted in an exception to this trend with greatly reduced supplies and relatively high retail prices.

Fresh Green Vegetables

The per head consumption of fresh green vegetables in 1987 was 310 grammes per week and of other fresh vegetables 518 grammes per week. The decline in consumption of fresh green vegetables of thirty per cent since 1965 follows from reductions in the per capita consumption of a number of the main individual items within the food group. For instance, average household consumption of cabbage has declined by nearly thirty per cent between 1965 and 1987 and the consumption of Brussels sprouts has also declined by some 55 per cent, as shown in Table 8.

Table 8. The Consumption of Fresh Vegetables

Grammes per person per week

	1965	1987
Cabbage	148.8	107.1
Brussels sprouts	82.2	37.0
Cauliflower	85.0	61.8
Leafy salads	38.3	50.7
Peas	33.9	11.8
Other fresh vegetables	6.9	11.2
Total	**442.9**	**310.1**

Source Household Food Consumption and Expenditure 1987, HMSO 1988

The household consumption of peas and beans has also gone down markedly and private gardens now contribute greatly to the supply. Some fresh green vegetables have also declined in supply because of improvements in freezing. The consumption of leafy salads, mainly lettuce, has gone up by twenty per cent since 1965. There has been a 60 per cent increase in the consumption of "other fresh vegetables" since 1965. This group comprises onions and root vegetables, tomatoes, cucumbers and mushrooms and a wide range of miscellaneous vegetables.

Table 9 shows that apart from the drought years of 1975 and 1976 there has been a steady rise in purchases of carrots, while consumption of turnips, swedes and other root vegetables is only a slightly lower than in 1965. There has been little change in the consumption of onions and tomatoes. Cucumbers and mushrooms, on the other hand, have shown an appreciable rise in consumption. Miscellaneous vegetables show nearly a 50 per cent increase in purchases over the last ten years and they reflect a wider choice. The proportion of households purchasing miscellaneous fresh vegetables in the survey week has doubled from ten per cent in the mid-sixties to over 20 per cent in 1987. It is most likely in the future that the consumption of salads will continue to rise, as also mushrooms, whose real price only rises slightly over time, and that sales of miscellaneous vegetables and salads will continue to increase.

Table 9. The Consumption of Other Fresh Vegetables

	1965	1975	1987
	grammes per person per week		
Carrots	98.0	84.2	125.5
Turnips, Swedes and other root vegetables	75.9	60.0	67.2
Onions Shallots and Leeks	96.4	90.7	105.1
Tomatoes	121.9	120.3	111.6
Cucumbers[a]		26.1	35.1
Mushrooms[a]		14.6	26.4
Miscellaneous[b]	53.5	32.7	47.6
TOTAL	**445.7**	**428.6**	**518.5**

a included under miscellaneous fresh vegetables in 1965
b including items such as celery, marrow, sweet corn and asparagus

Source: Household Food Consumption and Expenditure - 1989

The Market for Salads

The salad market in Britain comprises tomatoes, cucumbers and lettuce in the main, and celery, asparagus and minor commodities such as aubergines, peppers and Chinese leaves. Tomatoes dominate the group, with an annual apparent consumption per head of 7 kg. Lettuce is next in importance, having an annual consumption of over 4 kg per head, and cucumbers follow with an annual consumption of nearly 2 kg per head. The three salads are all characterised by large variations in seasonal consumption, the lowest in autumn and winter and the highest in spring and summer, but the distinct seasonal profile is in part determined by supply. The demand for salads is on the increase in Britain.[10]

There are two main points. Firstly, the demand for salads in Britain, and secondly, the competition in supply, taking into account supply from countries within the EEC, dominated by Holland, and that from third countries and their chances in this expanding market. Considering this competition, successful timing of supply is important, but a supply of quality produce is paramount. With regard to market shares, the salad market is dominated by tomatoes, at 54 per cent of consumer spending, followed by lettuce at 19 per cent, cucumbers at 14 per cent, spring onions (seldom imported) at five per cent, and finally radishes (also only token imports), watercress and beetroot, which together take five per cent of consumer spending. Celery takes three per cent of consumer spending. There has been a shift in the distribution pattern at the retail end. Supermarkets had 40 per cent of the sales, an

[10] The Market for Salads in Britain. Hinton Lynn. pp 87-111 Optimal Yield Management ed. Dan Rymon Avebury 1988.

increase of three per cent from 1983, and this has been lost in 1986 by traditional greengrocers, who now have 30 per cent. Market stalls remain steady in their share at 17 per cent.

The market remains dominated by the three most important salads: tomato, lettuce and cucumbers, and, some way behind, celery. Less popular salads now finding a foot-hold in the market are the exotics, mooli, radicchio and endive, as well as peppers, together with the multicoloured capiscum. In 1986 iceberg lettuce had doubled it share of the market from just over seven per cent in 1983. With its premium price and extra weight per piece, it now accounts for more than a quarter of all lettuce sales. Overall sales of lettuce have not expanded and iceberg has progressed at the expense of other crisp varieties, which have dropped their share of the lettuce market from 43 to 37 per cent since 1983. Even though beef and cherry tomatoes are more available now, 98 per cent of sales of tomatoes are still of the standard salad type.

EEC Policy and Salads

It has been mentioned that Spain and Portugal were disfavoured by EEC policy as compared with most Mediterranean countries, before their accession. For instance, the Common Customs Tariff on tomatoes was 11 per cent for Spain, but 4.4 per cent for Morocco, Algeria, Tunisia, Egypt, Cyprus and Malta, while the CCT on peppers was 6.3 per cent for Spain and 5.4 for Morocco, Algeria, Tunisia, Israel, Cyprus and Malta, with 4.5 per cent for Turkey. The result of Spanish and Portuguese accession will be to put them in a completely favourable position as members of the EEC.

The Outlook for Salads in Britain

The salad market in Britain has a most promising future. In the case of lettuce, consumption has increased greatly. Summer demand is mainly of interest to home producers but there is evidence of the emergence of a strong winter demand mainly supplied by imports from third countries with warmer climates. The demand is so great as to attract supplies from a number of distant countries because the winter price is so high in accordance with the strong demand. The outlook for lettuce is good since demand is increasing, bringing in more imports.

Tomatoes have not witnessed the same large shift in consumption as lettuce but 1983 to 1986 there was a distinct increase. At the same time imports have become a larger part of supply and this trend, evident since 1980, is likely to continue. Unlike lettuce, imports of tomatoes are dominated by a very few countries. These are Holland for the home season and the Canaries and Spain catering for the British off-season. While there is little scope for countries to compete with Holland in the summer, there is the opportunity for Mediterranean third countries to export some tomatoes in the off-season of home supply, and many of them have been doing this for some time.

Cucumbers have shown a bigger expansion in consumption than lettuce in recent years, and the proportion imported has increased markedly. Consumption, from a low point in 1973, has more than doubled and is now on a new rising trend. Again as with tomatoes two countries dominate imports, Holland in summer and the Canaries in winter. A number of Mediterranean third countries compete with the Canaries in the winter supply and they have scope for more exports to meet the rising winter demand.

Celery has increased in consumption only slightly over the years. For this commodity home suppliers dominate the market. The home-grown field crop is important and this shifts domestic supply further into the winter than the other salad produce. Imports do not really start till November and they come in strongly until May, when the home-grown crop takes over. In the off-season, imports are dominated by Spain and Israel and there is scope at this time for other Mediterranean third countries. The present trend in rising consumption gives every prospect of an expanding market in the winter period.

Britain is becoming more salad-conscious and home and imported supplies are working together to meet the rising market. Spending on salads, apart from their obvious importance to health, also depends on trends in disposable income. There is ample opportunity for third countries, and Mediterranean third countries in particular, to export salads to Britain in the off-season of home supply. The criteria are quality and freshness. The off-season prices are most attractive to exporters and will repay transport costs. In the home season of supply, exporting does not succeed because the ruling summer price is so low.

English growers can guarantee quality and freshness, as also can the north European exporters, especially Holland. Quality specifications, sometimes over and above those of the EEC, which are demanded by the supermarkets for English produce, must also be met by any exporter from whatever source. Consumption is expanding, but, the demand is not for quantity, though this will of course increase, but for quality, and any country to succeed in the export trade to Britain must be fully aware of this. The cold chain now ensures that the produce also arrives fresh. Freshness and quality are the key to the British market and competition on this basis will be intense. Health as a principle basis of demand, apart from the more sophisticated attributes of salad eating, is now a most important underlying factor in demand. A report of the Royal College of Physicians in 1986 cited Britain as one of the most unhealthy countries in Europe as regards diet. Exporters of salads can help Britain as well as helping themselves and home producers can benefit from a promising market.

While West Germany may be held up as the shining example in the consumption of health giving fruit and vegetables the large wholesalers in the UK are confident that the market will expand in Britain. With fruit and vegetables available as they now are in Britain there is no need for Britain to remain one of the most unhealthy countries in Europe so far as the fruit and vegetable sector is concerned.

Fruit and Vegetable Consumption

In the revised specification of the National Food Survey, income elasticity depend critically on the increase income level or the income share, at which the elasticities are evaluated.[11]
The income elasticities for a number of fruit and vegetables are given in Table 10.

Table 10. Income Elasticities of Certain Fruits and Vegetables

	Income Elasticity 1987		Income Elasticity 1988	
Fresh potatoes	-0.43	(0.08)	-0.53	(0.08)
Fresh green vegetables	0.37	(0.07)	0.24	(0.06)
Other fresh vegetables	0.35	(0.07)	0.30	(0.05)
Fresh fruit	0.66	(0.60)	0.53	(0.05)
of which:				
apples	0.43	(0.07)	0.47	(0.06)
oranges	0.21	(0.13)	0.28	(0.11)
Processed vegetables	-0.04	(0.05)	-0.02	(0.05)
of which:				
frozen peas	0.29	(0.12)	0.08	(0.14)
frozen chips and other frozen convenience foods	-0.08	(0.14)	-0.01	(0.14)
Other fruit and fruit products	0.77	(0.08)	0.55	(0.07)
of which:				
fruit juices	0.95	(0.18)	0.73	(0.10)

Standard error in parenthesis

Source: Household Food Consumption and Expenditure 1987 and 1988, HMSO 1989

---

[11] Household Food Consumption and Expenditure 1988, HMSO 1989

Response to changes in income is positive but inelastic (a coefficient of less than 1 at the income share used in evaluation) for all items presented in Table 10, although that for fruit juices in markedly higher than that for most foods. On the other hand, the coefficient for a number of foods is negative, indicating that they are "inferior goods". Less is consumed as incomes rise, at least at the income shares at which the coefficients are evaluated. It is important, in assessing the significance of the figures, that due attention should be given to the magnitude of the standard errors of the estimates which in turn are relatively large.

Fresh potatoes gave a negative income elasticity which suggests that fewer potatoes will be purchased as income rises. Fresh green vegetables, all fruit and the separate fruit items have positive income elasticities. There is little difference in the income elasticity for fresh green vegetables and for other fresh vegetables, and it is indicated that something in the order of one third or more will be purchased for one per cent growth of income. The income elasticity of all fresh fruit is 0.66 in 1987 which suggests that as income rises expenditure on fruit will increase by two thirds. The values though still positive are lower for apples at 0.43 and lower still for oranges at 0.21 (1987).

Table 11. Estimates of Price Elasticity of Demand for Fruit and Vegetables

	Estimates of price elasticity of demand			
	1987		1988	
Fresh potatoes	-0.14	(0.03)	-0.16	(0.05)
Fresh green vegetables	-0.39	(0.12)	-0.41	(0.13)
Other fresh vegetables	-0.24	(0.12)	-0.29	(0.13)
Fresh fruit	-0.53	(0.22)	-0.77	(0.23)
Processed vegetables	-0.20	(0.23)	-0.46	(0.22)
of which:				
frozen peas	-0.75	(0.34)	-0.87	(0.30)
frozen chips and other frozen convenience foods	-0.37	(0.22)	-0.45	(0.19)
Other fruit and vegetables	-0.67	(0.26)	-0.73	(0.30)
of which:				
fruit juices	-0.65	(0.23)	-0.80	(0.23)

Standard error in parenthesis

Source: Household Food Consumption and Expenditure 1987 and 1988, HMSO 1989

The estimates of own price elasticity of demand given in Table 11, have been derived from time-series analysis of the monthly survey averages of purchases and real (deflated) prices over the period 1982 to 1987. Many staple foods, such as potatoes and beans, have elasticities close to, or not significantly different from zero, reflecting a low response to price changes. Other foods such as certain frozen convenience foods including frozen vegetables have high (negative) elasticities with the coefficient sometimes greater than 1.

If the price falls by 1 per cent, potato purchases may be expected to be up by only 0.14 per cent in 1987. If the price of fresh green vegetables goes down by one per cent, purchases may be expected to rise by 0.39 per cent. Likewise other fresh vegetables purchases may be expected to rise by 0.24 per cent. Frozen peas, given a fall in price of one per cent, may expect to lead to an increase in purchases of 0.75 per cent, a much higher rise than for fresh vegetables. The response is less for frozen chips and other frozen convenience potato products, estimated price elasticity, -0.37 per cent, while the response of fresh fruit as a whole and for fruit juices have an estimated price elasticity of -0.67 per cent and -0.65 per cent.

The Food Survey also publishes estimates of cross-elasticities for oranges, apples and pears. The elasticity of oranges, apples and pears are given in Tables 12 and 13.

Table 12. The Elasticity with Respect to Price of Oranges, Apples and Pears, 1987

	Oranges		Apples		Pears	
Oranges	-1.38	(0.18)	-0.03	(0.13)	-0.12	(0.12)
Apples	-0.01	-0.05)	0.23	(0.09)	0.05	(0.05)
Pears	-0.32	(0.31)	0.30	(0.33)	-1.58	(0.43)

Source: Household Food Consumption and Expenditure 1987, HMSO 1988

Table 13. The Elasticity with Respect to Price of Oranges, Apples and Pears, 1988

	Oranges		Apples		Pears	
Oranges	-1.44	(0.20)	-0.22	(0.15)	-0.11	(0.12)
Apples	0.09	(0.06)	-0.19	(0.10)	0.05	(0.05)
Pears	-0.28	(0.30)	0.28	(0.32)	-1.70	(0.36)

Source: Household Food Consumption and Expenditure 1988, HMSO 1989

The Consumption and Purchase of Potatoes

The consumption and purchase of potatoes is given in Table 14, on a per week basis. The Table gives the weekly consumption per person in grammes for each quarter of the year which applies, and for the average week of the year. Consumption by type of potato is also given in kgs per head per year. Potato consumption is appreciably greater during the winter months then in the summer, and it is clear that the bulk of potatoes purchased is not pre-packed. On the basis of the Food Survey, the annual consumption of potatoes in Britain is 57 kg per head. The consumption in the UK is quite remarkable, but, in the longer term, consumption of potatoes in the fresh form will decline as it has done in the United States where over half the potatoes are processed in one form or another. In Table 15 both groups, fresh green vegetables and other vegetables show a similar pattern of consumption throughout the year. Consumption per head of all vegetables is 38.4 kg and the annual per head consumption of fresh green vegetables is 24.7 kg. During the first half of the year the per head weekly consumption of fresh green vegetables is below 300 grammes, while for July to September it is 392 grammes, and that for the last quarter of the year, is 332 grammes.

Cabbages of various kinds are bought by 25 per cent of the households in the survey week, and quarterly consumption of cabbage is more even than that for the group of "fresh green vegetables". This reflects the numerous types of cabbage which are harvested successively over the year. Consumption per head per year of cabbages is 4.5 kg. Leafy salads are by far the most popular type of vegetable in the fresh green vegetable group. Thirty-three per cent of households bought these vegetables in the survey week.

Table 14. Consumption and Purchase of Potatoes

	Consumption						Purchases	
	Jan/ March	April/ June	July/ Sept	Oct/ Dec	Yearly average	kg per year	Yearly average	kg per year
	grammes per person per week							
Old potatoes								
January-August								
not prepacked	827.98	583.50	67.49	-	369.82	19.23	357.69	18.60
prepacked	290.20	162.36	10.58	-	115.71	6.02	115.71	6.02
New potatoes								
January- August								
not prepacked	27.68	269.67	500.77	-	199.37	10.37	181.02	9.41
prepacked	4.35	45.72	96.11	-	36.70	1.91	36.70	1.91
Potatoes								
September-December								
not prepacked	-	-	315.08	993.76	327.21	17.01	298.59	15.53
prepacked	-	-	83.36	254.43	84.29	4.38	84.29	4.38
Total fresh potatoes	**1150.2**	**1061.2**	**1062.1**	**1248.1**	**1133.1**	**58.92**	**1074.3**	**55.86**

Source: Household Food Consumption and Expenditure 1988, HMSO 1989

Differences in seasonal demand for leafy salads in which lettuce is prominent, is most pronounced. Purchases increase greatly over the summer period. There is scarcely any consumption of leafy salads in the first quarter at 0.31 grammes per head per week. Consumption rises to 70 grammes per head per week in the April to June quarter and it rises further to 75 grammes per head per week from July to September. Consumption per head of this delicate type of vegetable is 2.5 kg per year. There is less waste with leafy salads than with the coarse vegetables. Cauliflowers, like cabbages, are harvested in Britain over the whole year. Imports of cauliflowers from Brittany are important in the coldest time of the year, from January to March. This is the quarter which has the lowest cauliflower consumption of 60 grammes per head per week. Cauliflower consumption increases in the spring quarter and in the summer period. From July to September cauliflower consumption reached 98 grammes per head, to fall off in the last quarter. Table 15 shows that 21 per cent of households bought cauliflower in the survey week, and consumption per head per year is 3.9 kg.

Most Brussels sprouts, as one would expect, are consumed in the October to December period, when consumption per person per week is 72 grammes. Consumption of Brussels sprouts from January to March is high at 6.3 grammes per head and in the spring and summer quarters it is notably less. Thirteen per cent of households bought Brussels sprouts in the survey week and yearly consumption per head is 1.7 kg. Peas and beans are minor vegetables in this group, with appreciably more consumption in the UK harvest season, than at other times of the year. The area under peas for market has been declining over the years and fresh peas in Britain are now relatively scarce.

The Consumption and Purchase of Other Fresh Vegetables

While fresh green vegetables in Table 15 comprise legumes and the brassicas including salads, the group of "other fresh vegetables" includes a large mixture of vegetable types, root vegetables and onions, the salad vegetables, tomatoes and cucumbers and mushrooms. Carrots are consumed most in the two winter quarters, a consumption of around 140 grammes per head per week. Though carrot consumption is lower in the spring and summer months it remains at upwards of 100 grammes per

Table 15. Consumption and Purchase of Fresh Green Vegetables and Other Fresh Vegetables.

	Consumption						Purchases		
	Jan/ March	April/ June	July/ Sept	Oct/ Dec	Yearly average	kg per year	Yearly average	kg per year	per cent house-holds*
	grammes per person per week								
Cabbages	93.93	109.17	101.09	99.84	101.09	5.26	87.09	4.53	25
Brussels sprouts	62.52	3.11	9.64	72.47	37.01	1.92	32.35	1.68	13
Cauliflowers	59.72	77.14	97.98	75.89	77.76	4.04	74.34	3.87	21
Leafy salads	0.31	69.98	75.27	32.97	52.56	2.73	48.21	2.51	33
Peas	4.98	7.78	18.35	9.95	10.26	0.53	2.80	0.15	1
Beans	16.17	17.73	78.07	19.91	32.97	1.71	8.09	0.57	3
Other fresh green vegetables	13.69	14.93	12.13	10.89	12.75	0.66	10.89	0.57	7
Total fresh green vegetables	282.11	299.84	392.53	322.23	324.10	16.85	263.45	13.70	63
Carrots	139.34	104.82	118.82	142.14	126.28	6.57	116.64	6.07	40
Turnips and swedes	44.79	27.37	19.60	49.14	35.15	1.83	32.04	1.67	11
Other root vegetables	30.48	10.58	17.11	32.04	22.71	1.18	18.35	0.95	9
Onions, shallots, leeks	111.66	102.33	99.22	113.22	106.69	5.55	97.04	5.05	38
Cucumbers	23.95	51.01	43.86	24.26	35.77	1.86	34.21	1.78	26
Mushrooms	30.48	27.99	28.30	28.62	28.93	1.50	28.62	1.49	28
Tomatoes	77.45	126.59	160.49	101.40	116.64	6.07	101.09	5.26	48
Misc. fresh vegetables	43.54	45.41	70.60	44.48	51.01	2.65	46.66	2.43	21
Total other fresh vegetables	501.70	496.41	558.00	534.67	522.85	27.19	474.95	24.70	78
Total vegetables	783.81	796.25	950.52	856.90	846.95	44.04	738.40	38.40	

* Per cent of households purchasing each type of food during the survey week

Source: Household Food Consumption and Expenditure 1988, HMSO 1989

person per week. Carrots are purchased by 60 per cent of the households in the survey week and carrot consumption is at 6 kg per head per year. This makes the carrot the most popular and generally used vegetable in Britain after the potato. Carrots are commonly the second vegetable on the menu. Turnips and swedes have an annual consumption of 1.6 kg per head, much lower than carrots. Like carrots, the uptake of turnips and swedes is higher in the winter months, when fewer non-root vegetables are available. Eleven per cent of the households bought turnips and swedes in the survey week. Other root vegetables, which together have a per capita consumption of one kilo, have the same seasonal pattern of supply as carrots but with a much lower supply, especially in the spring and summer.

The seasonal consumption of onions and similar vegetables is again similar to carrots, but consumption falls less in the spring and summer. Onions, shallots and leeks have an annual per capita consumption of 5 kg. Britain is some fifty per cent self-sufficient for onions but she depends on imports for the rest. Some fifteen to twenty overseas countries partake in the UK onion trade in various parts of the world. The Netherlands has the largest stake in the import market and traditional sources such as Poland and Egypt also export onions to Britain.

Cucumbers have an annual per capita consumption of 1.8 kg. During the first and last quarter of the year, the consumption of cucumbers is 24 grammes per person per week. This rises to 51 grammes from April to June and it is 44 grammes from July to September in line with summer demand. Twenty per cent of the households bought cucumbers in the survey week. Tomatoes enjoy a high annual consumption of 5.3 kg per head, and 48 per cent of the households buy tomatoes in the survey week. Seasonal consumption differences are pronounced. Tomatoes are taken up a 77 grammes per head per week from January to March, which is the season for the lowest supply. The supply of tomatoes increases to 127 grammes in the spring, and a 160 grammes in the summer. From this high level the consumption of tomatoes falls to 101 grammes in the last quarter of the year. The seasonal market profile for tomatoes is like that for cucumbers. Supply in the January to March quarter is largely from the Canary Islands, and from Spain. These tomatoes do not have the flavour and quality of the spring and summer supply but they are a most valuable supply over the winter. The Canary Islands are far enough south to grow tomatoes successfully in the winter, much to the advantage of countries in Northern Europe.

While tomatoes and cucumbers are available throughout the year there is more uptake of this produce in the spring and summer seasons when the demand for salads is at its peak. Nevertheless, ever since cucumbers and tomatoes have been grown commercially in the UK there has been a regular and substantial demand in the UK in what was formerly an "off-season" of supply, now shared in supply with Holland. This is the spring and early summer season.

In the case of mushrooms, advances in supply throughout the year have gone much further. Once a seasonal luxury, mushrooms have become an important feature of the household diet and while many consumers still think of mushrooms as being special, they have become an important and necessary ingredient in the household diet. Mushrooms are distinctive and without substitute. Annual mushroom consumption per capita has reached 1.5 kg per head, which is quite remarkable for a commodity which is so light in weight. Seasonal consumption of mushrooms is almost even throughout the year, at 28 grammes per person per week from April to December and 30 grammes in the next quarter. Mushrooms were bought by 26 per cent of households in the survey week.

The Consumption and Purchase of Fresh Fruit

Table 16 gives the consumption in grammes per person per week for a range of fruit. The Table also gives consumption in grammes for the average week of the year and annual consumption per head in kg. The apple is the fruit most commonly bought. Fifty per cent of the households bought apples in the survey week, and apples enjoy by far the highest demand at 10.5 kg per head per year. The consumption of apples over the year is surprisingly even at around 220 grammes per person per week. Demand is slightly higher in January to March quarter, the period when alternative fruit commodities are in relatively scarce supply. The demand for apples is a little less from July to September, when private garden supplies are available. The other staple fruit, bananas, are bought by thirty eight per cent of households in the survey week and annual consumption of bananas is 5.3 kg per head. Twenty-three per cent of the households bought oranges in the survey week. Quarterly consumption figures show oranges as being consumed most in the first half of the year, with appre-

Table 16. Consumption and Purchase of Fresh Fruit.

	Consumption						Purchases		
	Jan/ March	April/ June	July/ Sept	Oct/ Dec	Yearly average	kg per year	Yearly average	kg per year	per cent house- holds*
	grammes per person per week								
Oranges	128.77	129.70	79.00	51.94	97.35	5.06	97.35	5.06	23
Other citrus fruit	79.00	42.30	32.66	92.38	61.58	3.20	61.58	3.20	20
Apples	230.17	222.39	216.48	223.32	223.01	11.60	202.17	10.51	50
Pears	40.43	39.81	35.77	49.14	41.37	2.15	40.75	2.12	14
Stone fruit	8.40	24.26	108.24	4.67	36.39	1.89	35.15	1.83	11
Grapes	21.77	24.88	28.62	31.10	26.75	1.39	26.44	1.37	12
Other soft fruit	5.91	27.37	58.47	3.42	23.64	1.23	15.86	0.82	6
Bananas	95.18	115.71	127.52	107.31	111.35	5.79	111.35	5.79	38
Rhubarb	6.53	16.17	7.46	1.87	8.09	0.42	1.56	0.08	1
Other fresh fruit	10.26	23.33	39.19	23.64	24.26	1.26	23.95	1.25	7
Total fresh fruit	626.11	666.24	734.04	588.79	653.80	34.00	616.16	32.04	73

* Per cent of households purchasing each type of food during the survey week

Source: Household Food Consumption and Expenditure 1988, HMSO 1989

ciably less from July to September, and a big fall in the last quarter of the year.

Grapes which are among the most expensive of fruit have an annual consumption of 1.4 kg per head. While grapes are consumed more in the last quarter of the year when there is less fruit about, all quarters of the year show a high consumption which rises in each successive quarter. Grapes are at best a Mediterranean fruit and supplies formerly grown under glass in northern Europe are no longer available. All the grapes in the UK market come from such countries which have the right type of climate. To sustain the market for grapes in the UK for the whole year means that Southern hemisphere suppliers such as South Africa and New Zealand are strong in the market in Europe's off-season. The annual consumption of "other soft fruit" is 1.25 kg per head. Other soft fruit consumption is highest in the peak summer period, July to September at 58 grammes per head per week. Most of the consumption of "other soft fruit" in the quarters adjacent to the July to September quarter will actually be in the market only one or two weeks on either side of the July to September period.

Rhubarb is classed as a fruit because it is eaten as a fruit. Botanically it is a stem. There is a remarkably high consumption of rhubarb in the UK of 1.6 kg per year per person. The word "remarkable" is used partly as there is so little waste, but the answer to such high consumption is that there is little alternative fruit available when it is on the market in the peak months, April and May. Consumption is highest in its natural season in the UK, April to June with some imports from the European Mediterranean countries early in the season. The annual consumption of "other fruit", amounts to 1.3 kg per head. Consumption is most pronounced from July to September, which is similar to some other fruits. Consumption in this quarter is some two thirds of the yearly consumption and adjacent quarters have a one third consumption between them.

Overall, consumption of all fruit is highest from July to September at 734 grammes per head per week and consumption is next highest and in the previous three months. Consumption falls to 589 grammes in the October to December quarter. Despite seasonal differences in supply, the British public are well supplied with fruit over the year.

Fruit and Vegetable Consumption by Income Size Group

Data on consumption of fruit and vegetables by income group is given in Table 18, but to give perspective to the Food Survey results, the number of households interviewed and the number of persons represented is shown in Table 17 which gives the classification by income group and by region. So Table 17 discloses that the highest income groups, A1 and A2, relate to only 8.5 per cent of the households and to 10.4 per cent of persons. Group B, with incomes of between £205 and £375 per week, relates to the biggest share of the market, 29 per cent of households and 34 per cent of persons. Group C, with incomes of from £100 to £205 per week has 26.7 per cent of households and 29.7 per cent of the persons. Group D represents households with incomes of less than £100 per week and covers only 4.9 per cent of persons. The group of old age pensioners comprises 11.9 per cent of households and 6.4 per cent of persons. The purchasing power for fruit and vegetables is very limited in the groups which include old age pensioners and groups D and E2, both with incomes of £100 a week or less. These groups taken together account for a significant share of the households, at 30.2 per cent, and 22.0 per cent of persons.

The lower half of Table 17 gives the geographical background to the Food Survey, which is geared to the distribution of population over Great Britain. The figures for the number of households and the number of persons are closer together in this regional analysis. Eight five per cent of the households are in England, 5.3 per cent in Wales and 9.8 per cent in Scotland. Within England the largest share is in South East England and East Anglia, 31.3 per cent, the North West is the next region in prominence, with 12.2 per cent of households.

Table 17. The Number of Households and the Number of Persons Interviewed, by Income Group and Region

By Income Group

Income group	Weekly income of the head of the household	Number of households	Number of persons per cent
A1	£495 or more	226	3.8
A2	£375 and under £495	397	6.6
B	£205 and under £375	2121	34.0
C	£100 and under £205	1957	29.7
D	Less than £100	336	4.9
Total A1 to D		**5037**	
Households without an earner			
E1	£100 or more	411	3.8
E2	Less than £100	1002	10.7
OAP	n.a.	870	6.4
Total all households		**7320**	

By Region

	Scotland	717	9.6
	Wales	387	5.3
	England	6216	85.1
	North	457	6.3
	Yorkshire/Humberside	802	10.8
	North West	890	12.2
	East Midlands	513	6.9
	West Midlands	641	9.3
	South West	624	8.6
	South East/East Anglia	2289	31.0

Source: Household Food Consumption and Expenditure 1988, HMSO 1989

Vegetable Consumption in Britain

Fruit and vegetable consumption by income group is shown in Table 18. The consumption of all vegetables is important throughout the income classes, but more vegetables, as in West Germany are consumed by the lower income classes. Within this general trend the highest income group A1 shows the lowest consumption of vegetables. Even the next lower income group, A2 has an appreciably higher consumption than A1. Going down the scale of income, the amount of vegetables consumed increases all the way, from a weekly consumption of 2426 grammes per week for group B, to 3049 grammes for E1, the households without an earner but with an income of £100 per week or more. E1 is an interesting group. It is obviously a group intellectually endowed coupled with a good income endowment to appreciate more than any other social class the health value of vegetables.

Table 18. Fruit and Vegetable Consumption by Main Income Group for 1988

	Income group								
	Gross weekly income of household								
	Households with one earner or more						Households without an earner		
	£495 and over	£375 and under £495	£375 and over	£205 and under £375	£100 and under £205	Less than £100	£100 or more	Less than £100	OAP
Social class	A1	A2	All A	B	C	D	E1	E2	
	grammes per person per week								
Vegetables:									
Potatoes	752.08	857.21	821.13	999.36	1215.21	1383.17	1160.16	1400.59	1292.97
Fresh green	320.99	328.14	326.28	310.10	284.91	279.62	569.51	325.96	463.75
Other fresh	580.39	629.53	611.18	514.14	457.22	483.66	826.42	517.56	585.37
Frozen, incl. veg. products	168.89	251.63	221.46	209.02	220.21	154.58	169.51	198.13	143.70
Other processed, incl. veg. products	315.70	346.80	336.23	393.77	438.87	437.94	323.17	427.36	343.69
Total vegetables	2138.37	2413.94	2316.28	2426.07	2617.05	2738.97	3048.77	2869.61	2829.49
Fruit:									
Fresh	907.29	872.45	883.65	651.62	544.62	459.40	1200.60	562.97	777.59
Other, incl. fruit products	116.64	125.66	121.93	100.78	85.53	93.31	203.73	117.88	164.54
Fruit juices	478.37	406.52	433.27	268.73	185.69	111.66	282.42	149.92	117.57
Total fruit	1502.61	1404.63	1438.54	1022.68	815.84	664.68	1686.74	831.40	1060.01

Source: Household Food Consumption and Expenditure 1988, HMSO 1989

Groups E2 and the OAP group both have vegetable consumption appreciably above 2800 grammes per week, higher than that of other income classes except E1. This is largely due to the higher potato and processed vegetable consumption in E2, and higher consumption of fresh green vegetables and other fresh vegetables in the OAP class.

There are more differences between the social classes in the consumption of the different types of vegetables such that different income groups tend to consume different patterns of vegetables. A1, for instance, consumes outstandingly the least potatoes at 752 grammes per week as compared with the much heavier consumption of class D at nearly twice this level. This is no real surprise, but the magnitude of the difference is quite remarkable.

Persons in class A1 consume 321 grammes of fresh green vegetables per head per week as compared with 464 grammes in the OAP class, and a yet higher consumption of 570 grammes per week in the E1 class. Social class A1 consume 580 grammes per person per week of other fresh vegetables, E1 consumes 826 grammes per person per week and the A2 group consumes 630 grammes.

The amount of frozen vegetables, consumed by A1 is the second lowest consumption of all social classes, while A2 people consume more frozen vegetables than any other social class. Persons in group A1 are also the lowest consumers of processed vegetables, including vegetable products.

It is interesting to see that the group A2 has a higher consumption of processed vegetables than A1, of 347 grammes per head per week, but that this class of vegetable is consumed at a at a higher level by group B. Processed vegetables are known to lack the same health - giving attributes as fresh vegetables and they are a less enjoyable food. Processed vegetables are more relied on by social classes B and C. The consumption of processed vegetables is highest in group C, with a weekly income of £100 to £205, at 439 grammes per person per week. However, consumption of processed vegetables in group B is only fractionally lower.

At the same time, the class E1 with no earners and an income of £100 per week or more consume almost as little processed vegetables than the highest income group, A1, while the OAP group consume the least processed vegetables. The foregoing means that a large number of consumers prefer processed vegetables to fresh, most probably on the grounds of convenience and on price. This situation is a challenge to the fruit and vegetable trade and to the health lobby which is growing in Britain.

Fruit Consumption in Britain

The weekly consumption of fruit, except by the E1 group, the nature of which has been referred to, is highest in the highest income group A1 at 1503 grammes per head per week, with a relatively small drop in A2. Consumers in group B take only two thirds of the quantity of fruit consumed by the A groups, or 1023 grammes per head per week. Consumption continues to fall in groups C and D to some half the level of the A groups.

Somewhat curiously the OAP group consume a little of fruit more than those in group B and even households with no earner and an income of less than £100 per week, eat a little more fruit than group C. Only three components make up fruit consumption shown in Table 18, and the differences in consumption between the social classes are just as great as for vegetables. Group E1, which is again on its own, has much the highest consumption of fresh fruit at 1200 grammes per head per week. Apart from E1, A1 persons consume the most fresh fruit at 907 grammes and the amount falls with the declining scale of income to 652 grammes for group B, 545 grammes for group C and 459 grammes for group D. E2 has a consumption of fresh fruit of 563 grammes but the OAP group consumes more than B at 778 grammes per head per week. Fresh fruit dominates the fruit supply, but fruit juice consumption is more important in the higher income groups. Group A1 takes 478 grammes of fruit juice per week per person and A2 takes 407 grammes. People in group B consume 269 grammes. In Group C consumption is 186 grammes per person, Group D consumes 112 grammes. Group E2 consumes 150 grammes, and the OAP group consumes 118 grammes per person per week. Fruit juice, weight for weight, is more important than fresh fruit because of its high concentration. Of course any steps which provide for a more equal income structure and raising of real income would lead to a notable expansion of the demand for fruit and vegetables.

The consumption of other fresh fruit, including fruit products, suggests that on the whole less is consumed as incomes fall. Groups A1 and A2, taken together, consume 122 grammes per week. This falls to 101 grammes in group B and to 86 grammes in group C, but in D consumption rises to 93 grammes, while in E2 consumption rises again to 118 grammes. In the OAP class consumption rises much more to 165 grammes. Overall, OAP's consume comparatively more vegetables than fruit, and those with less than £100 per week, either earned or unearned, consume the least amount of fruit, especially in the D1 group, which is comparatively better off for vegetables. The descending profile of consumption, going down the scale of income, falls more steeply in fruit than in vegetables. This shows that income is more a factor determining higher consumption in fruit than it is for vegetables. This is particularly true since fruit as a whole is much more expensive than vegetables as a whole.

Fruit and Vegetable Expenditure

Expenditure on Vegetables

Fruit and vegetable expenditure in terms of pence per person per week by income group is given in Table 19 for the same social groups and with the same breakdown of fruit and vegetables given in Table 18.

The two class A groups, together, spend more on vegetables than any other social class with the exception of the special class E1. Expenditure on vegetables falls in all other social groups as income falls. This means that the more affluent buy higher priced, better quality vegetables and more expensive salad vegetables than the less affluent. This follows because, in Table 19, it is clear that the amount spent generally increases as incomes decline, even though the magnitude of the change is not very great. Potatoes, for instance, have expenditure for A1 at 75 per cent of that for E2. It is evident that better quality potatoes are bought by the higher income groups. The two A groups spend much more on fresh green vegetables than all the other groups, again with the exception of E1, which has the highest expenditure of income groups including A1. Expenditure on fresh green vegetables declines throughout the scale descending income down to and including D. The E2 and the OAP group each spend 539 pence and 722 pence respectively on fresh green vegetables and each of these figures is substantially above the D group with an expenditure of 462 pence. Fresh green vegetable expenditure is much higher in the more affluent classes. It is 872 pence in A1, 582 pence in B but 462 pence in D. The other fresh vegetable class attracts less expenditure throughout the scale of diminishing income over the groups from A1 to C inclusive. Group D spends a little more on fresh green vegetables than do B, E1 is the usual exception with the highest expenditure of all, but both E2 and the OAP group spend somewhat more on fresh green vegetables than class D.

Expenditure on frozen vegetables including vegetable products is highest in the A2 group, at 715 pence per week, and next highest in the B group at 589 pence per week. It is almost as high in group C, but expenditure on frozen vegetables including vegetable products falls to 408 pence a week for D. It is 524 pence per week for group E1, 490 pence for group E2 and 403 pence for the OAP group. Expenditure in groups B and C is almost identical, but this drops substantially in D. In E2, the class with no earner with an income of less than £100 per week, frozen vegetables, including vegetable products, show expenditure is significantly up on group D, but expenditure by the OAP group is virtually the same as in D. Incomes in these categories are likely to be less than £100 a week, the first two by definition. Expenditure on other processed vegetables including vegetable products is highest in group C with earnings of £100 to £205 a week and almost identical to group B which is the next higher income bracket. This is at the level of some 1500 pence per person per week. Expenditure on other processed vegetables including vegetable products is 1402 pence for all A, but it drops substantially to 1265 pence in D and, typically, for E1 it is down to 1063 pence per person per week. E2 spends more than D and the OAP class has the lowest expenditure of all, at 976 pence. The health seeking class E1 group avoid processed vegetables.

Expenditure on Fruit

Again, the range of expenditure on fruit as for vegetables is greater than the range in consumption and there is a lack of symmetry, unlike vegetables, in the trend of income and expenditure over the income classes. To some extent the latter must be due to the larger assortment that is presented by fruit. The social class A1 has the highest expenditure on fruit as a whole and A2 follows A1 in this. There is a big difference between the all A group at 3973 pence and the B group at 2617 pence, and

Table 19. Fruit and Vegetable Expenditure by Main Income Group for 1988

	Income group								
	Gross weekly income of household								
	Households with one earner or more						Households without an earner		
	£495 and over	£375 and under £495	£375 and over	£205 and under £375	£100 and under £205	Less than £100	£100 or more	Less than £100	OAP
Social class	A1	A2	All A	B	C	D	E1	E2	
	pence per person per week								
Vegetables:									
Potatoes	616.78	602.16	609.94	639.49	686.45	788.78	722.53	811.80	789.41
Fresh green	871.83	733.73	787.23	583.19	488.32	461.89	1065.29	538.71	721.91
Other fresh	1852.52	1728.42	1774.77	1339.01	1076.49	1090.18	1873.99	1135.28	1134.34
Frozen, incl. veg. products	499.21	715.07	636.69	588.79	578.84	408.39	524.09	490.19	402.79
Other processed, incl. veg. products	1375.71	1408.37	1402.46	1507.28	1512.25	1264.67	1063.43	1298.88	976.34
Total vegetables	5215.43	5187.75	5210.77	4657.75	4342.36	4014.22	5249.34	4274.87	4024.79
Fruit:									
Fresh	2747.37	2301.04	2463.40	1653.46	1322.83	1079.91	2968.83	1338.38	1766.06
Other, incl. fruit products	597.81	578.84	580.08	411.50	324.41	316.63	816.47	455.04	556.44
Fruit juices	1097.95	832.02	929.68	552.09	370.13	192.53	581.32	296.42	232.03
Total fruit	4443.13	3711.89	3972.85	2616.74	2017.68	1589.70	4366.31	2089.84	2554.53

Source: Household Food Consumption and Expenditure 1988, HMSO 1989

another significant drop to 2018 pence in group C and another one to 1590 pence for group D. Class E1 shows expenditure nearly up to that of A1. E2 with an unearned income below £100 per week spends more than group C, with a weekly income of £100 to £205. The surprise is that the OAP group spends so much on fruit, almost as much as B with a much higher income.

Fresh fruit expenditure is highest for the most affluent group A1 and it declines with income falls to and including group D, and leaving group E1 alone, expenditure on fresh fruit increases from 1080 pence in D, to 1338 pence in E2, while in the OAP group it increases to 1761. The special class, class E, has an expenditure of almost 3000 pence, 200 pence or so above A1. The OAP group would seem to value fruit and vegetables in much the same way as E2.

Other fruit, including fruit products is most popular, as one would expect amongst the A groups at a 580 pence per week. There is a big drop in purchases in group B at 411 pence and further drops in C and D to 317 pence per week. As usual E1 is special with an expenditure in other fruit including fruit products of 816 pence per week, much higher than A1 at 598 pence per week. Expenditure in E2 at 455 pence is greater than that in group D at 317 pence, and also greater than that in groups B and C. The OAP class has expenditure of 556 pence on other fruit which is close to that in A1 and A2 combined. This also shows how the OAP group value fruit.

Fruit juice expenditure shows a very big range, from 1098 pence in A1 to 192 in group D. There is a big drop from all A at 930 pence to B at 5.52, to C at 370 pence and to D where expenditure is little more than half that in C. The special group E1 spends more than class D by a large margin and the OAP group exceeds class D in the consumption of fruit juice.

Fruit and Vegetable Consumption in the UK Regional Deviation from the National Average

There are marked deviations in the distribution of fruit and vegetables over Britain. These are given in Table 20.

Table 20. Regional Deviation from the National Average of Fruit and Vegetable Consumption in the UK

	Fresh fruit	Other fruit and fruit products	Fresh potatoes	Fresh green vegetables	Other fresh green vegetables
	percentage deviation from the national average				
North	-16	-19	11	-11	6
Yorks and Humberside	-9	-24	4	- 4	-6
North West	-13	-15	7	-20	-9
East Midlands	- 5	-10	10	12	-6
West Midlands	- 8	-14	10	3	-1
South West	13	11	-4	30	5
South East and East Anglia	18	26	-14	14	8
Wales	-10	-12	13	-3	3
Scotland	-22	-14	7	-47	-15

Source: Household Food Consumption and Expenditure 1987, HMSO 1988

The region best placed for "fresh fruit" purchases is the South East of England group East Anglia and the South West. Scotland is the most disadvantaged region for fresh fruit. The same applies to Northern England and the North West, which are disadvantaged to a lesser degree. But even Yorkshire and Humberside is disadvantaged in the supply of fresh fruit, as is Wales and the East and

West Midlands. This situation, striking as it is for the UK, must be repeated, perhaps to a lesser extent in other countries in the EEC.

For "other fruit and fruit products" those regions disadvantaged for the distribution of fresh fruit, are in the main disadvantaged in the distribution of fruit and fruit products. The South East and East Anglia and the South West are advantaged as before, particularly the South East for which there is good access to imported and home supplies and where the better incomes give the fruit and vegetable trade a sounder commercial basis in which to operate.

In the distribution of fresh potatoes the South East and East Anglia is disadvantaged. Most advantaged are Wales, the East Midlands and the West Midlands and the North. The potato trade is different from the point of view of supply. Potatoes are durable and subject to long term trade connections and the advantaged regions are close to a source of potato production. This explains the positive deviation in Scotland.

"Fresh grown vegetables" illustrate extreme deviations in distribution and consumption. The South West is richly served with a +30 and the East Midlands have a +12, while the South East and East Anglia region again come out well. Scotland is very strongly disadvantaged in the distribution of fresh green vegetables, while the North West is also disadvantaged, more than it is for fruit. The North, strongly disadvantaged for fruit is also disadvantaged for fresh green vegetables.

For other fresh green vegetables, Scotland, as one might expect, is the most disadvantaged region. Yorkshire and Humberside and the North West and the East Midlands are all disadvantaged in the supply of other fresh green vegetables. This is the situation revealed by the Food Survey. It would be of benefit if the challenge to the trade, which the situation discloses, were to be taken up at least in some areas and in some fruit and vegetable produce. There are good margins to those in the trade for those who want to do something about it.

Table 21. The Consumption of Fresh Fruit

	1965	1975	1987	1988
	grammes per person per week			
Oranges	99	107	84	97
Other Citrus	35	46	66	62
Apples	229	211	219	223
Pears	24	23	33	41
Stone fruit	17	11	47	36
Grapes	a	10	21	27
Soft fruit	40	18	24	24
Bananas	111	89	100	111
Rhubarb	b	16	11	8
Misc. fresh fruit	29	14	25	24
Total	584	545	630	654

[a] included in the total for soft fruit
[b] included in the total for misc. fresh fruit

Source: Household Food Consumption and Expenditure 1988, HMSO 1989

The Consumption of Fresh Fruit in the UK

A profile of the consumption of fresh fruit from 1965 to 1988 is given in Table 21. Oranges are seen to have reached a saturation point, orange consumption is up and down with the seasons but shows a level and large consumption overall. Other citrus by contrast are on a rising trend, from 35 grammes per week in 1965 to over 60. The consumption of apples is remarkably consistent level at a high level of consumption, while the consumption of pears, the poor relation, shows an upward movement. There is also evidence of a remarkable increase in the consumption of stone fruit, the great majority of which is imported. Grapes show a sustained and large increase in consumption and soft fruit a substantial increase, but not as large. Banana consumption repeats the story of oranges and apples, a remarkably steady and high consumption over the last two decades or so. Rhubarb, commercially a fruit, seems to be doubling in consumption, while that of miscellaneous fruit is increasing.

Overall fresh fruit consumption is seen to be on a distinct upward trend confirmed by the sustained upward level in 1987 and 1988. This is 11 per cent up on 1965.

CHAPTER 6

THE NORTH TEMPERATE EXPORTER OF FRUIT AND VEGETABLES : THE NETHERLANDS

The Dutch Export Market

Introduction

The Netherlands or the low countries became the economic centre of northern Europe in the Middle Ages because of favourable geographic factors. The present State of the Netherlands is also known in the United Kingdom as Holland. History has also resulted in a low-lying county in England being given the same name. In fact Holland is one of the many Dutch provinces or separate fiefs which originally owed allegiance to the Kings of France, or to the Holy Roman Empire. In this Chapter both titles are used but the Netherlands is of course much more accurate and the Dutch have used it since gaining nationhood early in the nineteenth century after centuries of struggle.

This chapter relies heavily on the valued data from the Central Bureau of Dutch Auctions. The considerable Dutch exports of fruit and vegetables may be summarised as follows. In 1988 the Dutch Auctions handled 1160 million kg of glasshouse vegetables worth 2020 million D fl. It handled 800 million kg of outdoor vegetables worth 680 million Dfl. The Auction also handled 380 million kg of fruit worth 390 million Dfl, in addition to 50 million kg of mushrooms worth 140 million Dfl. In all produce exported amounted to 2390 million kg worth 3240 million D fl.[1]

Table 1. Dutch Vegetable Exports by Country of Destination - million kg.

West Germany	800	Scandinavia	113
United Kingdom	264	Switzerland	27
France	168	Austria	12
Belgium-Luxembourg	167	United States	7
Denmark	34	Canada	1
Italy	20	Ireland	18
Iceland	2	Saudi Arabia	3
Africa	65	Canary Islands	1
Spain	22	Others	52

Source: Central Bureau of Dutch Auctions 1989

[1] From a Summary of Dutch Exports of Fruit and Vegetables - The Central Bureau of Dutch Auctions 1989.

Table 2. The Netherlands - Vegetable Production 1987, Areas and Quantities

Items as listed in Eurostat	'000 ha	'000 tonnes
Cauliflowers	2.3	43.9
Brussels sprouts	5.9	87.2
Cabbage	1.8	114.8
Other brassicas	1.6	59.2
All brassicas	**11.6**	**305.1**
Celeriac and celery	1.1	49.5
Leeks	2.8	75.8
Lettuce	3.5	125.8
Endive	0.8	41.2
Spinach	1.6	50.5
Asparagus	2.8	10.3
Chicory	4.8	70.1
Other vegetables (leafed or stalked)	-	19.0
Vegetables leafed or stalked	**17.4**	**442.2**
Tomatoes	2.3	547.2
Cucumbers	1.4	357.0
Gherkins	0.6	38.0
Melons	-	3.6
Aubergines	0.1	20.1
Pumpkins, marrows	-	3.5
Peppers	0.5	66.3
Cultivated for fruit	**4.9**	**1035.7**
Kohl-rabi	0.1	3.9
Turnips	-	9.3
Carrots	3.9	260.7
Garlic		
Onions	15.5	478.2
Beetroot	0.4	21.7
Salsify and scorzonera	0.8	19.2
Radish	-	32.7
Other roots and tubers	-	1.3
Roots and tubers	**20.7**	**827.0**
Peas	6.4	27.3
French beans and runner beans	5.3	50.6
Other pulses	1.1	10.9
Pulses	**12.8**	**88.8**
Cultivated mushrooms	**0.1**	**113.0**
Other fresh vegetables	**3.4**	**15.0**
Total of vegetables specified	**70.9**	**2826.8**

Source: Eurostat 1989

The production base for these exports is as follows. The area of vegetables under glass in 1988 was 4323 ha. That of vegetables grown in the open was 47595 ha, of fruit 24362 ha, and mushrooms 92 ha. Eighty per cent of the glasshouse vegetables are exported, 42 per cent of the outdoor vegetables, 36 per cent of the fruit and 22 per cent of the mushrooms. Over 56 per cent of all fruit and vegetables grown in Holland is put on the export market. The export volume of fresh vegetables grown in Holland, 1776 million kg in 1988, to the various countries including the distant markets of USA and Canada is shown in Table 1.

The Dutch Vegetable Area and Production

The area and production of vegetables in the Netherlands from Eurostat data for 1987 is given in Table 2. The Netherlands has a total vegetable area of 71 thousand ha, which covers 3.5 per cent of the used agricultural land. This compares with some 180 thousand ha in the UK. The total output of vegetables in Holland is 2.8 million tonnes. Of the total area of vegetables, 11.6 thousand ha is under crops of the brassica type, which yield 0.3 million tonnes in volume, and 17.4 thousand ha are under leafed or stalked vegetables, which account for a production of over 0.4 million tonnes. Within the 17.4 thousand ha of leafed and stalked vegetables, the most prominent vegetable is chicory, which covers 4.8 thousand ha and yields a production of 70 thousand tonnes. Lettuce, with an area of 3.5 thousand ha, has a production of 126 thousand tonnes, and asparagus with 2.8 thousand ha has a production of 10 thousand tonnes. Salad vegetables are prominent in the outdoor vegetable cropping in the Netherlands and in the Dutch export trade.

Next there is an important sector of vegetables "grown for fruit". The sector includes important export crops, tomatoes, cucumbers and gherkins, which like peppers are grown under glass. Tomatoes cover 2.3 thousand ha and give a production of virtually 0.55 million tonnes. Cucumbers, which cover 1.4 thousand ha, have a production of 0.36 million tonnes. The less prominent vegetables, melons, aubergines, pumpkins, marrows and peppers are all substantially grown under glass. The range of crops in the vegetables cultivated for fruit group is large for a north temperate country, such crops are tomatoes and cucumbers and a proportion of the lettuce, but this is much as may be expected from Holland, because the Netherlands is the leading vegetable grower throughout the world.

The group of roots and tubers cover the largest area of all the vegetable groups, 20 thousand ha with a production of 0.83 million tonnes. Much the largest area, 15.5 thousand ha or three quarters are down to onions. Onion production is 0.48 million tonnes, much of which is sold on the export market and all of which is grown to a very high standard. Carrots, at 3.9 thousand ha, and with a production of 0.26 million tonnes are the only other vegetable in this group with an area exceeding one thousand ha. Carrots, again, are grown to perfection and a proportion is exported. Of course the vegetable group which has the highest proportion of exports are those vegetables "cultivated for fruit", which like all Dutch production of fruit and vegetables is grown to a high standard. The reputation of the Dutch in this field of activity is known to all as second to none. Leguminous vegetables make up the balance of the area with beans and peas occupying 5.3 thousand ha and 6.4 thousand ha respectively, with a production of 51 thousand tonnes and 27 thousand tonnes. The area down to leguminous vegetables is high in the Netherlands and the production of green peas for market has not declined to the extent as it has done in the UK.

The Dutch Fruit Area and Production

The area and production of fruit in the Netherlands from Eurostat data for 1987 is given in Table 3. The Netherlands grows 24 thousand ha of fruit which compares with the UK fruit area of 49 thousand ha. Fruit production in Holland accounts for 1.2 per cent of the usable agricultural land. The entire production of fruit is 0.52 million tonnes. There is a very small production of grapes, which is an anachronism relating to a former age, when the production of hot-house grapes, like the more extensive area in Belgium, was well known.

The Netherlands grow 20 thousand ha of apples and pears which account for eighty per cent of the Dutch fruit production in area. These two commodities together have a production of 0.48 million tonnes, which accounts for 92 per cent of the total fruit production in the Netherlands. In the UK apples and pears are only 63 per cent of the total fruit area, while in the Dutch sector pears are more prominent than they are in the UK.

Stone fruit in the Netherlands accounts for only five per cent of the total fruit area, or 1.3 thousand ha. The production of stone fruit is eleven thousand tonnes. A great range of soft fruit is grown, though this sector in the Netherlands does not feature so greatly as it does in the UK. The area of soft fruit in the Netherlands is 2.7 thousand ha, 2.2 of which is strawberries. This gives rise to a production of soft fruit in the Netherlands of 29 thousand tonnes, 25 thousand of which is strawberries. The strawberry crop in the UK occupies 6.4 thousand ha out of a total soft fruit area of 14.5 thousand ha.

Table 3. The Netherlands - Fruit Production 1987, Areas and Quantities

Items as listed in Eurostat

	Area '000 ha	Production '000 tonnes
Fresh grapes		**0.6**
Apples	15.1	340.0
Pears	5.2	140.0
Apples and pears	**20.3**	**480.0**
Cherries	0.6	2.8
Plums	0.6	8.4
Other stone fruit	0.1	
Stone fruit	**1.3**	**11.2**
Black and red currants	0.2	2.5
Strawberries	2.2	25.0
Raspberries		0.3
Gooseberries		0.2
Other Soft fruit	0.3	1.1
Soft Fruit	**2.7**	**29.1**
Total of fruit specified	**24.3**	**520.9**

Source: Eurostat 1989

Dutch Exports of Vegetables

Table 4 and Figure 1 give the Dutch exports of non-salad vegetables and Table 5 and Figure 2 those of salad vegetables by country of destination. The pattern of vegetable exports given in Figures 1 and 2, show that West Germany takes the largest share of the Dutch exports of salad vegetables. The figures at the end of the Table 5 give the total vegetable exports for 1988, 1987 and 1986 and the per cent of these exports going to each country. Total exports of vegetables for Holland have risen progressively over the three years 1986 to 1988. Those to other members of the EEC, at 86.8 per cent of Dutch vegetable exports in 1988 have gone down relatively in the three years given, but the proportion going to other countries in the EEC is well over four fifths. This substantial proportion may decline a little as Spanish exports increase. Exports to other European countries were six per cent of all

Table 4. Dutch Exports of Vegetables, excluding Salad Vegetables, by Country of Destination - tonnes

	TOTAL	WG	BL	UK	FR	DE	IT	IR	SP	EEC	PER CENT	OTHER COUNTRIES	PER CENT
Asparagus	4917	4522	274	1	1	21	5	-	-	4824	0.7	93	0.0
Aubergines	18088	6013	643	4496	5119	411	1	88	12	16783	2.3	1305	0.2
Cauliflower	2540	958	82	7	136	102	155	7	71	1518	0.2	1022	0.1
Curly Kale	293	102	167	-	-	-	-	-	-	269	0.0	24	0.0
Spring Carrots	4600	2179	1977	3	123	214	-	2	3	4501	0.6	99	0.0
Maincrop carrots	55378	16351	26249	2975	7710	608	345	676	5	54919	7.6	459	0.1
Maincrop carrots/washed	89296	11363	69184	4016	3363	239	50	217	5	88437	12.2	859	0.1
Broccoli	1080	575	118	57	49	60	1	9	8	877	0.1	203	0.0
Chinese Cabbage	3626	386	332	430	180	949	7	38	66	2388	0.3	1238	0.2
Chinese vegetables	507	56	-	113	314	19	-	-	1	503	0.1	4	0.0
Courgettes	2286	1367	173	102	16	157	-	77	1	1893	0.3	393	0.1
Peas	5473	4	5465	-	-	3	-	-	-	5472	0.8	1	0.0
Celeriac	28062	12804	2496	165	7179	1028	2508	14	65	26259	3.6	1803	0.2
Kohl rabi	2047	1515	246	137	9	1	-	-	-	1908	0.3	139	0.0
Swedes	302	124	3	1	46	-	-	-	-	174	0.0	128	0.0
Beetroot	5696	2564	235	57	1111	247	583	1	35	4833	0.7	863	0.1
Paprika	58134	22121	1431	15912	447	1865	50	918	2	42746	5.9	15388	2.1
Parsnips	44	17	1	4	-	1	1	-	-	24	0.0	20	0.0
Sugar peas	33	9	5	8	1	1	-	2	3	29	0.0	4	0.0
Leeks	25921	13117	1238	2278	2836	914	591	85	103	21162	2.9	4759	0.7
Runner beans	928	148	518	229	9	2	-	-	-	906	0.1	22	0.0
Turnip tops	24	23	1	-	-	-	-	-	-	24	0.0	-	0.0
Onions	377136	75540	41774	39335	68652	7477	281	5675	16307	255041	35.2	122095	16.9
Spring onions	11295	2746	3461	1447	1605	42	223	-	1014	10538	1.5	757	0.1
Mushrooms	25600	8722	2340	9282	808	336	-	-	66	21554	3.0	4046	0.6
Cut vegetables	386	346	34	1	5	-	-	-	-	386	0.1	-	0.0
Other vegetables	59	32	1	9	1	3	7	1	-	54	0.0	5	0.0
TOTAL	723751	183704	158448	81065	99720	14700	4808	7810	17767	568022	78.5	155729	21.5
Percent specified country of total		25.4	21.9	11.2	13.8	2.0	0.7	1.1	2.5	78.5	.0	21.5	

Note: Portugal - only small amounts, included in "Other countries"

Source: Central Bureau of Dutch Auctions, 1989

Figure 1. Dutch Exports of Vegetables, Excluding Salad Vegetables

Exports by Quantity - 1988

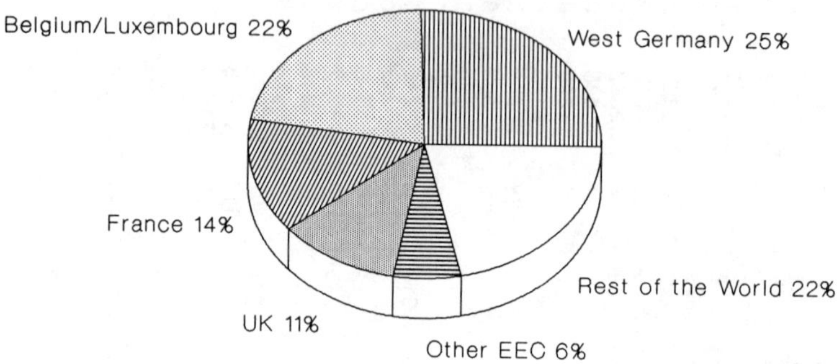

Source: Central Bureau of Dutch Auctions, 1989

Exports by Value - 1988

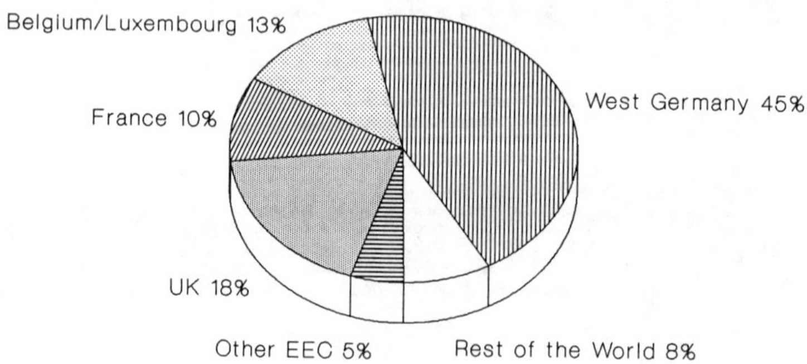

Source: Eurostat External Trade Exports, 1990

Table 5. Dutch Exports of Salad Vegetables by Country of Destination, 1988 - tonnes

	TOTAL	WG	BL	UK	FR	DE	IT	IR	SP	EEC	PER CENT	OTHER COUNTRIES	PER CENT
Gherkins	23997	20423	451	1092	472	428	14	-	125	23005	2.5	992	0.1
Celery	982	290	6	26	24	73	173	4	151	747	0.1	235	0.0
Paksoi	249	73	6	90	35	-	1	2	-	207	0.0	42	0.0
Lettuce	81698	63315	1147	7319	943	1073	1002	245	609	75653	8.2	6045	0.7
Chervil	140	1	139	-	-	-	-	-	-	140	0.0	-	-
Fennel	2912	725	198	429	1040	31	48	17	21	2509	0.3	403	0.0
Cucumbers	282408	214788	1914	25829	17031	10305	26	386	6	270285	29.4	12123	1.3
Mini cucumbers	266	129	-	120	9	1	-	-	-	259	0.0	7	0.0
Peppers	338	151	5	98	10	17	-	5	-	286	0.0	52	0.0
Endive	3048	1462	582	35	646	11	-	4	1	2741	0.3	307	0.0
Parsley	119	26	39	4	4	2	-	-	3	78	0.0	41	0.0
Purslain	58	-	58	-	-	-	-	-	-	58	0.0	-	-
Round Tomatoes	376008	203548	356	107975	23201	5967	406	6032	65	347550	37.8	28458	3.1
Beef tomatoes	108210	29366	279	5655	47778	235	9090	118	1417	93938	10.2	14272	1.6
Christmas tomatoes	1138	432	60	360	54	42	1	11	1	961	0.1	177	-
Garden cress	1410	1324	1	1	-	1	-	-	-	1327	0.1	83	-
Radish	36607	23318	2068	3786	3759	778	550	45	286	34590	3.8	2017	0.2
TOTAL	919588	559371	7309	152819	95006	18964	11311	6869	2685	854334	92.9	65254	7.1
Percent specified country of total		60.8	0.8	16.6	10.3	2.1	1.2	0.7	0.3	92.9		7.1	

Total of All Fresh Vegetables, 1988,1987 and 1986 - tonnes

											OTHER EUROPE	OTHER COUNTRIES
1988	1838217	814788	200822	262967	223723	34506	20244	17886	22359	1597295	111170	129752
Per cent	100.0	44.3	10.9	14.3	12.2	1.9	1.1	1.0	1.2	86.9	6.0	7.1
1987	1805084	827398	187820	273880	214958	35393	9756	20844	5116	1575165	120132	109787
Per cent	100.0	45.8	10.4	15.2	11.9	2.0	0.5	1.2	0.3	87.3	6.7	6.1
1986	1725360	833078	166546	276402	205310	29166	10020	22163	908	1543593	89850	91917
Per cent	100.0	48.3	9.7	16.0	11.9	1.7	0.6	1.3	0.1	89.5	5.2	5.3

Note: Portugal - only small amounts, included in "Other countries"

Source: Central Bureau of Dutch Auctions, 1989

Figure 2. Dutch Exports of Salad Vegetables

Exports by Quantity - 1988

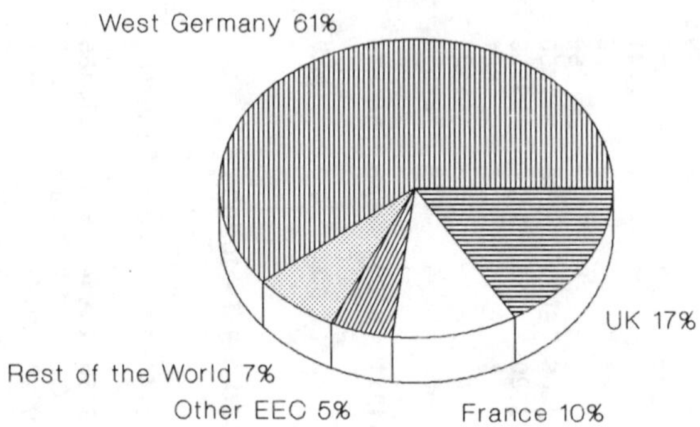

Source: Central Bureau of Dutch Auctions, 1989

Exports by Value - 1988

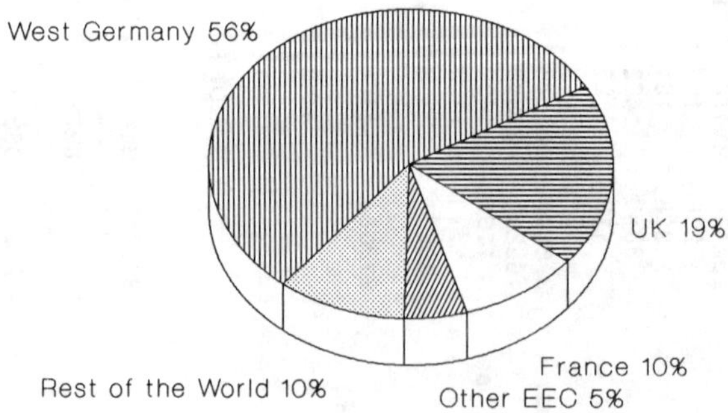

Source: Eurostat External Trade, Exports, 1990

Dutch vegetable exports in 1988, 6.7 per cent in 1987 and 5,2 per cent in 1986, and the proportion exported to "other countries", though on a rising trend, is similar to the proportion exported to other European countries. It is the Dutch vegetable exports to these countries, which are rising, which has the effect of lowering slightly the proportion of vegetable exports to the EEC. In the EEC, West Germany takes by far the most imports of any single country, 44.3 per cent in 1988, 45.7 per cent in 1987 and 48.2 per cent in 1986. Next comes the United Kingdom, some way behind West Germany but appreciably ahead of other EEC members except France. The United Kingdom's share of Dutch vegetable exports was 14.3 per cent in 1988, 15.1 per cent in 1987 and 16.0 per cent in 1986. Belgium-Luxembourg for the size of population is a very prominent importer but proximity to Holland is also a factor. Belgium-Luxembourg took 10.9 per cent of Dutch exports of vegetables in 1988, 10.4 per cent in 1987 and 9.7 per cent in 1986. Thus Belgium-Luxembourg shows a rising trend in the export market for Dutch vegetables, against the trend for the EEC as a whole. France took 12.2 per cent of Dutch vegetable exports in 1988, 12.1 per cent in 1987 and 11.9 per cent in 1986, showing a very slight increase over the three years.

Denmark imports 1.9 per cent of Dutch exports of vegetables, fewer than her population of five million would suggest. In 1987 Denmark imported two per cent of the Dutch exports of vegetables and in 1986, 1.7 per cent. In 1988 the Republic of Ireland imported one per cent of Dutch vegetable exports. This shows a decline in 1988 from 1.2 per cent in 1987 and 1,3 per cent in 1986. It is unlikely that the exports to Ireland will decline further because the level of Dutch imports of vegetables has already reached such a low plane. Italy and Spain are both relatively self-sufficient for vegetables. Italy took only 0.5 per cent of Dutch vegetable exports in 1987 and only 0.6 per cent in 1986, but Dutch exports of vegetables to Italy went up to 2.1 per cent in 1988. Dutch exports of vegetables to Spain have grown perceptively over the years 1986 to 1988 (Spain joined the European Community in January 1986) but Dutch vegetable exports to Italy and Spain will never be a significant amount because of the large domestic production of vegetables in these two countries. The small exports of Dutch vegetables to Italy and Spain in fact represent the interests of consumers in these two countries to add to their range of an already varied and abundant supply.

As has been mentioned, Belgium-Luxembourg, the United Kingdom and France are important markets for Dutch vegetables, but Dutch exports of vegetables to West Germany exceed the combined vegetable exports to these countries.

Table 5 shows the Dutch exports of salad vegetables which illustrates certain preferences for salad vegetables in the countries concerned. For instance, West Germany takes 85 per cent of the Dutch exports of gherkins, 78 per cent of the lettuce, 76 per cent of the cucumbers, 64 per cent of the radishes and 54 per cent of the round tomatoes. This also corresponds with West Germany's comparatively large lack of self-sufficiency in this produce. West Germany and Holland are in strong competition in the West German market. For several generations now Dutch produce in many lines has been put on the West German market. Until World War 2 when there was one Germany and a larger market at competitive prices and a high level of quality. A larger proportion of the high value salad vegetables as opposed to other vegetables exported from Holland find a market in the EEC which takes 93 per cent of the Dutch exports of salad vegetables. Of the total vegetable exports from Holland, 87 per cent go to the EEC member states who depend on Holland for a wide range of salad vegetables, while the type and amount of vegetables imported by individual member states depends on their own degree of self-sufficiency in the various types of vegetables. Put another way, without Holland the demand for salad produce in the EEC member countries would not be met. The same goes for other vegetables exported from Holland. For "other vegetables", Belgium-Luxembourg is seen to be greatly dependent on Dutch exports. For instance, Belgium-Luxembourg also takes well over half the Dutch exports of main-crop carrots, and almost as many spring carrots as the more populated West Germany. Belgium-Luxembourg take almost the entire smaller export of fresh peas and over half the runner beans. Every country in the EEC takes some vegetable exports from Holland, without which the consumer satisfaction in the EEC would be considerably reduced.

"Other European countries" specified in the schedule of Dutch vegetable exports are Switzerland, Austria, Norway, Sweden, Finland and Iceland. "Other countries" feature the United States, Canada, Africa and Canary Islands and countries not listed separately. Of these the largest export market for Dutch vegetables is Saudi Arabia, followed some way behind by the United States.

Table 6. Dutch Exports of Fruit by Country of Destination - 1988 tonnes

	TOTAL	WG	BL	UK	FR	DE	IT	IR	SP	EEC	PER CENT	OTHER COUNTRIES	PER CENT
Grapes	234	104	1	9	1	3	57	2	-	177	75.6	57	24.4
Apples	81622	45348	5976	11324	6807	2109	5	1348	224	73141	89.6	8481	10.4
Pears	62599	9680	1244	25879	6093	4793	3094	868	811	52462	83.8	10137	16.2
Cherries	120	7	112	-	-	-	-	-	-	119	99.2	1	0.8
Plums	10	7	-	-	-	-	-	-	-	7	70.0	3	30.0
Melons	618	268	3	293	-	7	-	11	-	582	94.2	36	5.8
Strawberries	8581	4379	395	2504	486	287	89	16	9	8165	95.2	416	4.8
Raspberries	39	23	6	1	1	4	-	-	-	35	89.7	4	10.3
Blackcurrants	106	92	-	11	-	1	2	-	-	106	100.0	-	-
Redcurrants	59	38	3	7	7	1	1	-	1	58	98.3	1	1.7
Gooseberries	13	10	-	-	1	-	-	-	-	11	84.6	2	15.4
Blackberries	100	63	-	5	1	23	-	-	-	92	92.0	8	8.0
Bilberries	272	167	23	31	2	15	20	-	-	258	94.9	14	5.1
TOTAL	154373	60186	7763	40064	13399	7243	3268	2245	1045	135213	87.6	19160	12.4
Percent specified country of total		39.0	5.0	26.0	8.7	4.7	2.1	1.5	0.7	87.6		12.4	

Note: Portugal - only small amounts, included in "Other countries"

Source: Central Bureau of Dutch Auctions, 1989

Figure 3. Dutch Exports of Fruit

Exports by Quantity - 1988

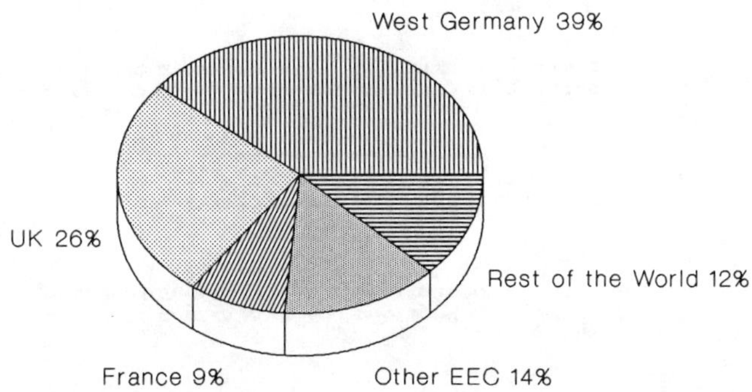

Source: Central Bureau of Dutch Auctions, 1989

Exports by Value - 1988

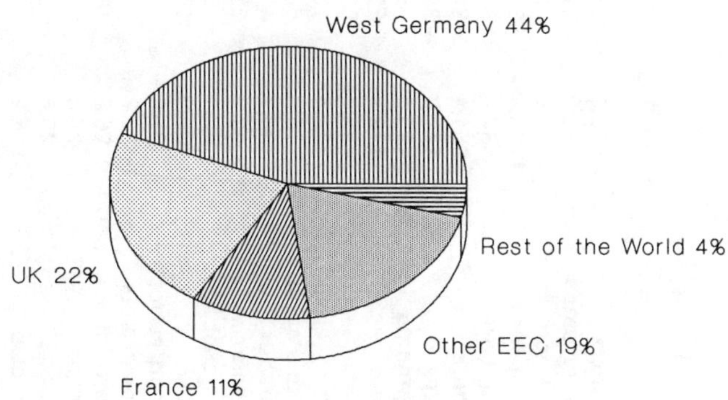

Source: Eurostat External Trade, Exports, 1990

Table 7. Dutch Exports of Fruit and Vegetables by Value - Million Dutch Florins, 1988, in descending order of value.

	TOTAL	WG	UK	FR	BL	SC a	AU b	SW c	US	CANADA	OTHER COUNTRIES
Tomatoes	878.9	468.3	249.4	54.6	0.9	60.3	7.1	21.6			16.7
Beef tomatoes	252.2	67.7	13.2	112.4	0.7	20.1	3.3	9.0	4.3	0.7	21.5
Cherry tomatoes	2.6	0.9	0.9		0.2	0.3		0.3			
All tomatoes	1134.4	536.9	263.5	167.0	1.8	80.7	10.4	30.9	4.3	0.7	38.2
Cucumbers	484.6	351.8	55.8	25.8	3.4	39.5	0.9	6.0	0.5		0.9
Peppers	251.1	82.9	74.0	1.8	4.8	45.7	2.3	3.3	27.7	4.4	4.2
Lettuce	177.7	135.3	19.6	1.9	1.5	7.8	6.2	0.8			4.6
Mushrooms	125.3	45.0	44.8	4.7	7.8	12.9	6.7	0.5			2.9
Apples	112.9	63.7	15.6	11.5	6.7	8.2					7.2
Chicory	93.7	35.0	2.0	11.9	16.5	0.8	2.3	15.5	0.5	0.5	8.7
Pears	85.4	13.6	34.7	9.3	1.4	12.6					13.8
Brussels sprouts	75.8	53.0	1.4	11.0	0.5	1.0	0.2	5.5			3.2
Asparagus	45.7	41.8			2.1	0.7	0.4	0.1			0.6
Strawberries	43.9	21.2	14.6	2.6	1.1	2.8	0.8				0.8
White cabbage	42.1	9.3	17.6	5.9	1.2	3.6	0.1				4.4
Leeks	38.9	16.2	4.3	4.2	1.2	8.6	0.7	1.2			2.5
Washed carrots	37.2	8.3	4.1	1.8	22.2	0.3		0.3			0.2
Winter carrots	30.7	10.9	3.0	4.1	8.4	3.1	0.2	0.2			0.8
Iceberg lettuce	26.6	15.1	2.5	0.2	0.3	4.4					4.1
Gherkins	24.3	19.4	0.8	1.3	0.8	1.9		0.1			
Celeriac	20.8	9.0	0.3	5.1	1.8	1.7	0.2	0.7			2.0
Cauliflower	5.2	1.6		0.2	0.2	1.5					1.7
Other produce	289.1	116.2	23.5	42.1	76.6	15.6	2.1	3.2	1.0		8.8
TOTAL	3145.4	1586.2	582.1	312.4	160.3	253.4	33.5	68.3	34.0	5.6	109.6

a= Scandanavia b= Austria c= Switzerland

Source: Central Bureau of Dutch Auctions, 1989

Dutch Exports of Fruit

The exports of fruit from Holland, shown in Table 6 and Figure 3, give a somewhat different picture from those of vegetables. In particular, only three fruits, strawberries, apples and pears, account for much the most of the Dutch fruit exports. In 1988 as much as 87.6 per cent of Dutch fruit exports went to countries in the EEC, in 1987, 80.2 per cent and in 1986, 82.9 per cent. This is very much the same situation as for Dutch exports of vegetables. The pattern for the overall fruit exports to members of the EEC follows their interest in acquiring the three main fruits mentioned. Thus exports of fruit to West Germany were 38.9 per cent of all Dutch exports in 1988, 44.5 per cent in 1987 and 33.8 per cent for 1986. The fluctuations in the proportion of exports to West Germany substantially reflects the level of the West German production of apples and pears in a particular year. West Germany has a large apple and pear sector. In general the West German import of fruit from Holland is much lower than that of vegetables. Belgium-Luxembourg imports of fruit from Holland were five per cent in both 1988 and 1987, and 6.9 per cent in 1986. The United Kingdom is, comparatively, a big importer of fruit in most years, again depending on the level of her own production of apples and pears. In 1988 Britain imported 26 per cent of Dutch fruit, in 1987, 13.9 per cent and in 1986, 19.3 per cent. In the three years France took a range of Dutch imports of fruit from 8.7 to 18.8 per cent, so indicating great variations in the French production of apples and pears. All EEC members take some fruit from Holland which also may fluctuate from season to season. "Other Europe" took from 8.8 per cent to 13.1 per cent over the three years and the Dutch exports of fruit to "other countries" ranged from 3.6 to 6.7 per cent over the three years 1986 to 1988.

Dutch Exports of Fruit and Vegetables by Value

Dutch exports of fruit and vegetables by value given in Table 7 put salad vegetables at the head of the list. Tomatoes by value alone exceed one third of the exports of all vegetables and with cucumbers, peppers and lettuce, salad vegetables amount to no less than 65 per cent of the Dutch exports of vegetables by value. On the fruit side, apples pears and strawberries, account for almost eight per cent of fruit and vegetable exports by value. West Germany takes fifty per cent of the Dutch exports by value, the United Kingdom 18.5 per cent and France 9.9 per cent. Belgium-Luxembourg take five per cent and other countries less than four per cent. It is a tribute to the Dutch fruit and vegetable industry that the Bureau of Auctions give so much data on fruit and vegetable exports. There is one other feature of the Bureau of Auctions data, which is fruit and vegetable exports per cent in each month, as given in Table 8.

Round and beef tomatoes cover the north temperate season of glasshouse production from March to the end of the year and they are in full production from April to October. Cucumbers are exported over the same season, peaking from April to September inclusive. Peppers have the same seasonal pattern but they peak from May to October inclusive. Lettuce exports are geared highly to the winter months but exports of lettuce feature all the year round. Mushrooms, not surprisingly, show exports all the year round with no particular peak.

Apple exports increase in July and August when the harvest season starts, but these exports also increase over the winter storage season as sales take place from store. Chicory corresponds with lettuce with all year round exports which are heavier in the winter season. Brussels sprouts have a pronounced winter season while asparagus has a short season, mainly May to June. Strawberries are mainly exported in June and July, but there is evidence that Dutch growers also cater for a late season market, in August and September, which is also a feature of UK production. White cabbage is exported all the year round but exports are somewhat higher when white cabbage is taken out of store, from March to June. Leeks are exported every month of the year, but more so in the months from September to December inclusive. Carrots also are exported throughout the year and the winter supply from store is higher than that in the summer months.

The export of Iceberg lettuce is quite seasonal with more exports in the summer months, the reverse of lettuce cited earlier. Gherkins are only available for export in the summer, peaking in July and August, while celeriac, though available all the year round, is exported most in the winter season. Exports of cauliflower take place from when the production season starts in April. Cauliflower exports peak from July to September inclusive to fall with the onset of frost, towards the end of the year. From the market export pattern of the Dutch fruit and vegetables, it is quite clear that the Dutch do not just export fruit and vegetables, but they target their exports over a long a season as possible. At the same time, they export strongly in the natural harvest season of a number of fruit and vegetables.

Table 8. Dutch Exports of Fruit and Vegetables 1988, per cent supply each month

	JAN	FEB	MAR	APR	MAY	JUN	JUL	AUG	SEP	OCT	NOV	DEC
Tomatoes	-	-	2	8	19	19	18	16	9	7	2	-
Beef tomatoes	-	-	3	9	18	15	15	15	11	10	3	1
Cherry tomatoes	-	-	1	5	9	11	16	24	19	12	3	-
All tomatoes	-	-	3	9	20	18	16	15	9	8	2	-
Cucumbers	-	2	9	11	17	16	15	12	10	7	1	-
Peppers	-	-	2	7	12	14	14	18	13	14	5	1
Lettuce	8	10	19	14	10	5	4	5	4	5	6	10
Mushrooms	8	8	10	8	9	8	8	9	7	9	8	8
Apples	10	12	13	7	8	3	1	6	8	16	10	6
Chicory	13	13	15	9	7	2	2	2	2	9	11	15
Pears	8	9	9	6	5	2	0	4	17	19	12	9
Brussels sprouts	12	4	1	-	-	-	-	2	10	25	24	22
Asparagus	-	-	-	3	58	38	1	-	-	-	-	-
Strawberries	-	-	-	2	9	31	39	12	4	3	-	-
White cabbage	6	7	11	9	12	9	5	8	7	9	8	9
Leeks	9	6	4	6	9	4	5	9	10	14	12	12
Washed carrots	5	7	10	9	8	3	3	9	12	13	15	6
Winter carrots	13	11	13	7	6	3	2	5	8	12	11	9
Iceberg lettuce	1	-	2	5	11	16	14	15	11	15	5	5
Gherkins	-	-	-	-	5	11	24	49	12	-	-	-
Celeriac	14	14	8	3	2	2	2	5	7	11	16	16
Cauliflower	-	-	-	1	5	12	19	18	19	17	8	1

Source: Bureau of Dutch Auctions, 1989

The Dutch Auction System of Fruit and Vegetable Marketing

We are told that on July 29th 1887 Dirk Jongerling in the north of Holland came by a boat filled with cauliflowers to the market place to find a buyer. The cauliflower supply was small and in order to save time-consuming negotiation he called out a high price and gradually brought it down. The first buyer to offer an accepted bid acquired the cauliflowers. This is stated to be the basis of the Dutch auction system. From 1887 the auction system for fruit and vegetable marketing developed rapidly in Holland. The Dutch auction system of selling has in fact, been used in the Netherlands since 1594 not for fruit and vegetables, but for fish.

Today in the Dutch Auction System a clock is used with a hand that is calibrated to swing from the highest to the lowest price. The clock represents a package of measures which together form marketing policy. Auctions cover many produce specialisations which are grouped under the Central Bureau of Fruit and Vegetable Auctions in the Netherlands. The Auctions belong to and exist for the growers, who believe in the principle and the benefits of cooperation. Producers put their entire effort into producing top quality produce on their farms, while the auction organisation is mainly involved in planning and promoting sales.

This unique auction system has now existed in Holland for over 100 years. The auction system, where supply and demand meet face to face is one of the reasons for the prominence of Holland in fruit and vegetable exports mainly on a European basis, and of ornamentals on a world basis. A stimulus for the auction system in Holland has been the national and official obligation to sell though the auction in the periods of crisis or wars, in the periods 1916-1919 and 1934 to 1964.

In his paper prepared for the centenary of the Dutch auction given in 1987, Barendse estimates the nature of fruit and vegetable sales by the collective system and the Dutch auction system for European countries, as shown in Table 9.[2]

The table shows that the Dutch auction system of selling is important in four European countries, Netherlands, Belgium, Sweden and Denmark. In Denmark, West Germany, France and Italy collective sales are based on an obligation by the public authorities. In Spain and the United Kingdom individual sales predominate.

The sale of fruit and vegetable products in Dutch horticultural sector are uniform. The producer grows and harvests produce and prepares it for sale. The producer grades and packs his product, and at this point it is taken up by the trade, when the auction assembles produce for sale, after every consignment is inspected by the auction staff. The auctions as cooperative societies are responsible for transferring the ownership of these products from the producer to the trade.

Table 9. Collective and Dutch Auction Systems of Marketing Fruit and Vegetables

Country	Collective Sales per cent	Dutch Auction Sales per cent
The Netherlands	98	90
Belgium	75	60
Sweden	60	50
Denmark	80	25
West Germany	45	10
France	40	10
Italy	40	2
Spain	10	2
The United Kingdom	20	-
Poland	40	-

Source: Conditions for a Successful Introduction to the Auction System in Horticulture, Barendse H.F.J., 1988

The Nature of Horticultural Produce

The basic condition of a successful Dutch auction is the nature of the produce. If the produce is not perishable the system will not function. For perishable horticultural products, like strawberries

[2] The Conditions for a Successful Introduction to the Auction System in Horticulture. Barendse H.F.J. Acta Horticulturae No 223 pp 17-23 1988

or summer apples (which do not keep) the individual producer has hardly any influence on the harvest day or the day the produce needs to be sold. In the case of stored apples however, the producer or the sales cooperative have the possibility to postpone the date of sale without risk of quality.

The establishment of the Dutch Auction for fruit and vegetables in the Netherlands was most likely related to confidence. In the 1880s there was a severe crisis in agriculture in many countries and following improvements in transport there was a switch in the Netherlands towards horticultural products. Over the same time the effects of the industrial revolution increased demand all over Europe for what were then luxury foods. Once the auctions started it needed to prove its usefulness. The auction provides transport and price formation. It also facilitates fast and efficient dispatch and it assures the producer of payment. At the same time it gives the opportunity to improve quality control. A strong and developed banking system, as part of the Auction business, provides credit for producers' investment and adds to the viability of the auction as a business. This has not been very important at times when the Dutch government made cheap credit available to producers. From time to time cheap credit for the fruit and vegetable sector is one of the aims of the Dutch government policy in this sector.

The Dutch auction system mostly occurs in a region of concentrated production and an important part of this production is sent some distance away. It is true that general sales problems or crisis periods have often led to financial support, but the willingness of producers and the belief in the cooperative principle by them is of course an essential element in the foundation of Dutch auctions.

In the auction room, ranks of buyers sitting in a tiered formation, see the number and description of the lot going through the auction. As the dial falls the buyer presses a button to make his bid. Also each buyer has a phone so he can acquire full knowledge of the prices in the main markets throughout Europe - say Cologne, Brussels, Rungis and Munich. The system is unique in how the price is fixed. The demand and supply situation of economic theory is wholly reflected and price fixing is instant. The system has been evolved over the years and it now depends on quality classification, standardisation packaging and air conditioning. The use of computers and modern communication systems has become commonplace. At virtually all auctions, now, the traditional manual clock has given way to computerised clocks, making tele-auctioning with other auctions possible. The purpose of the clock and auction system is to achieve the best possible price for the grower through direct confrontation between supply and demand, in public. The auction does not only match demand with supply, but for each sale the buyer is guaranteed produce which is reliable, which will keep and which can be offered on the international market. Dutch produce is circumscribed by measures such as quality control specified by the Auction and the Dutch government. This is associated with air conditioning, and standardised produce enables the buyer to buy on description. The block system, in which produce of a particular quality, grade and colour is sent in by growers, combines the smaller lots into one large consignment, which also adds to the efficiency of the system. Of course, the block system represents an advantage to the grower too, as small-scale producers are able to offer their produce as large-scale units which may then go to a distant market and fetch a higher price than otherwise.

Without a cooperative attitude on the part of the affiliated growers, it would be impossible to maintain the auction and the system behind it. The market discipline is imposed by the growers themselves and the producers own the auction. The auctions are autonomous, but in order to standardise sales and national measures, the Central Bureau of fruit and Vegetable Auctions in the Netherlands (CBT) was formed as far back as 1917. The CBT belongs to the auctions which in turn belongs to the producers.

Today the "Garden of Europe", a description aptly given to the Netherlands, offers over a hundred different varieties of fruit and vegetables. The grower accepts the quality standards and imposes them on his own premises where he selects and packages his produce according to these strict standards. For instance, for produce in Class 1, the "Dutch girl" logo is attached to or placed inside the packaging as a guarantee symbol. A Class 1 super designation is given to the best produce in the Class 1 range. Produce in Class 1 may be packaged and exported in non-returnable boxes. Incoming produce is inspected by the auction's personnel, supervised by controllers from the Quality Control Bureau. Also there is the psychological pressure that no grower wishes to see his produce downgraded at the auction, in front of his fellow producers. Moreover, it is not allowed for the grower to jeopardise the reputation of produce in a block. It is the evolution of the fruit and vegetable auctions, closely controlled by the producers but also controlling the growers produce, which has made the Netherlands produce pre-eminent in Europe and unrivalled by any markets in the world. 'Quality' is the name of the game and through the auction system quality is guaranteed. Dutch produce, distributed in many countries in Europe is sought by consumers because it has earned a reputation for quality which is second to none.

It is of some interest to note that when the European Common Market formulated policy for the fruit and vegetable sector in the early 1960s, when there were only six members of which the Netherlands was one, the quality standards were based on those which had already existed in Holland for some decades. No other EEC member country had in any way developed quality control as had the Netherlands. Furthermore, no member country has gained as much as Holland has done in the European market for fruit and vegetables through strict standardisation, quality control and high production per hectare. It goes without saying that production has to be perfect to grow produce of the right quality in the first place and the Dutch producers have always supported the auction by sending first class produce. Should sub-standard produce arise there are regulation safeguards built in to the system which avoids its sale.

Dutch Export-Orientated Horticulture

In a recent paper Meijaard discusses the export sector of Dutch horticulture and its progress. Horticulture in developed countries and especially the export-oriented sector, depends strongly on other branches of economic activity. These include supply and services, product treatment and processing and the trade and distribution sectors. Horticultural products generate not only producers' incomes but employment for workers in horticulture. They also give employment to workers in the service industry, including processing. The significance for the national economy of these horticultural groups in terms of income and net contribution to the balance of payments is considerable.[3] In Table 10 the net contribution is expressed as a per cent of the total of the horticultural sector, and the export dependent income as a per cent of the total income of the complex.

Table 10. The Net Contribution to the Balance of Payments

	Net contribution per cent		Export dependent income per cent
Flowers and pot plants	42		74
Vegetables under glass	21		72
Horticulture under glass		**63**	
Vegetables in the open	14		52
Flower bulbs	10		71
Fruit	9		66
Nursery stock	4		74
Horticulture in the open		**37**	
		100	

Source: The Dutch Agricultural Research Institute 1987.

[3] Characteristics of Export-Oriental Horticulture and its Developments. Meijaard D. Acta Horticultae No 223 pp 28-34 1987

Practically all agribusiness complexes are strongly dependent on foreign demand. The share of export dependent income, in the total income, amounts to 64 per cent for the total agribusiness. For the horticultural sector the percentage is even higher. Vegetables grown in the open rank low, fruit ranks intermediate and other branches, glass grown vegetables and ornamentals, rank high. Table 10 shows the net contribution to the balance of payments as a percentage of the total horticultural sector and the export dependent income as a percentage of the total income of the sector. This table excludes distribution and capital assets.

The position of other businesses in the open are less favourable. Production of open ground horticultural produce is strongly affected by weather at growing and harvest. Opportunity for export from this sector is to a greater extent determined by periodic shortage on the export market. To exploit these regularly occurring shortages the nursery stock sector grows a broad variety and the open ground fruit and vegetable sector stores a big proportion of its produce under modern storage conditions. The economic advantage of storage at the point of production is, however, becoming less as storage facilities in neighbouring countries are created. Also there is now a year round availability of fresh produce from other parts of the world.

Competition, Profitability and Investment

The wholesale markets of Western Europe bring produce from all parts of the world. Competition is intense and the interdependency of production and trade firms is accordingly great. Changes in production techniques or distribution have led to adjustments in home production. This strong mutual dependency presses on profit margins such that firms just remain in business. The yearly balance of savings are used as a yardstick for assessment of the future of the business. Meijaard considered that yearly savings should be at least 2.5 per cent of the value of business assets. As seen in Table 11, the bulb growing, pot plant and nursery stock sectors are the only ones which meet this condition.

The validity of Meijaard's factor, however, depends in practise on the profitability of the produce and the degree of capital investment concerned, hence the low factors for mushroom production should be adjusted for capital intensity. Also, that horticultural businesses continue in production at ratios below 2.5 calls into question the appropriateness of the ratio chosen.

Table 11. Savings as a per cent of the assets of specialised firms - four year average 1982-5

Vegetables under glass	-0.5	Flower bulbs	3.8
Cut flowers under glass	1.2	Fruit	2.0
Pot plants	3.3	Vegetables in open	1.1
Mushrooms	0.7		

Source: The Dutch Agricultural Research Institute 1988.

However, around this ratio there is a great range between businesses which show considerable dynamism. Small and obsolete businesses disappear continuously while the better firms take over their business and new firms are founded. In the past ten years the number of glasshouse businesses in the Netherlands decreased from thirteen to eleven thousand, while the area under glass increased from eight to nine thousand ha. An identical tendency is seen in the mushroom sector, a decrease from 900 to 800 firms but an increase in the area of mushroom production of 100,000 metre^2. The conclusion is that forceful competition leads to moderate profitability with regular adjustments in management to maintain profit at this level. This is equally true of horticultural enterprises in the UK.[4] Investigations in the Dutch trade sector give comparable results. The trend for increasing size of business and reduction in their number will continue. Family farms predominate and such firms are often combined with others. Each partner contributes his knowledge, capital and labour into the new business and economic viability is strengthened. The new owners are mostly content with a lower reward than may be achieved outside horticulture, and their living expenses are comparatively low. This can be illustrated by figures and results of firms in the glasshouse industry.

Table 12. Savings as a per cent of the Value of the Assets of Specialised Glasshouse Businesses and Age of Entrepreneur

Type of business	Age in years	Savings per cent of assets
Single proprietor firms	<35	4.5
Single proprietor firms	35-45	1.0
Single proprietor firms and a successor	>45	0.0
Single proprietor firms with no successor	>45	0.5
Firms with:-		
More proprietors, father and son(s)		8.0
More proprietors other than above		9.0

Source: The Dutch Agricultural Research Institute 1987

The willingness to cooperate in management for improving competition improves the profitability of the firms and creates changes for continuity of the business. In Holland young entrepreneurs, as a rule, are better equipped with management qualities than the older generation.

The Origin of the Self-sustaining Enterprises in Horticulture

The basis for the position of Dutch horticulture in international trade was set up at the turn of the century. Industrialisation in the surrounding countries was coupled with a strong urbanisation and an increase in incomes.[5] The demand for fruit and vegetables increased and the domestic production of these industrialising countries was not able to meet this increased demand. The Dutch growers and trade firms took their opportunity, a position which they still hold. At the turn of the century Dutch international trade in horticultural produce began to function. Dutch produce became supplemented by produce of foreign origin. Gradually the season of Dutch export trade was extended and broadened so that the trade connection could be brought about. More recent changes in the distribution of fruit and vegetables, involving supermarkets, substantially changed the trade structure, and has led to an increase in the scale of purchasing. Because of the willingness of the Dutch growers to cooperate in the auction system, exporters were able to satisfy their needs of mass distribution and delivery of quality graded uniform produce with continuity and successful handling of complaints.

Geographic centres of production emerged. Examples of these is the glasshouse industry in the Westland, the Aalsmeer complex, the bulb growing complex along the western part of Holland,

---

[4] Economic Results from Horticulture, Annual series. Hinton Lynn and Housden Walter, Department of Land Economy, University of Cambridge. Series

[5] Van Soest. A Century of Trade in Fruit and Vegetables. 100 years of Horticultural Produce Auctions 55-79 1987

and Boskoop as the centre of nursery stock production. The outcome is that the financial results in the areas of concentration are more favourable than elsewhere.[6]

In these centres the social communication structure, and the growers themselves, through their qualities attitudes and behaviour, make optimum use of resources. New technical knowledge, and its application, spreads easily, so that the viability of the business through a high flight of management is maintained. Centres even have their own horticultural weekly newspaper. Earlier mutual support took place through family connections, friends and neighbours. Now, specialist study clubs in the various sectors have developed. Separate study groups relate to the cultivation of a particular crop. Frequent excursions are made by growers to individual holdings, while technical and financial records are presented and management problems are discussed. Members in the group do not see each other as competitive. Defares puts forward that such interdependence is based on the idea that the exchange of knowledge and help in solving the problems of others, which can in future also be profitable for oneself.[7] The effect is to continually raise management and performance standards throughout the group as a whole. Such economic strength is not common in Western Europe at this level, though similar groups are present among the specialised locations in Belgium and towards the Dutch border in West Germany where the organisation of the horticultural sector is similar to the Dutch.

A parallel may be drawn between these developments, as "self-sustaining" horticultural groups, and the increase of sound management. In the sixties the accent was on craftsmanship, the knowledge and resources to attain a high standard of product. Craftsmanship or perhaps "plantmanship" still remains a condition for profitability. In the seventies, however, attention turned to additional factors in profitability such as management and entrepreneurship. In this context, the involvement as to what happens in the market and the eagerness to explore new possibilities are important. Abellas draws attention to the fact that a relatively important number of entrepreneurs in the glasshouse sector say that besides having economic goals, they also have social psychological goals, including good relations with colleagues, collaborators and advisory services.[8]

Weighing negative and positive effects of information technology, it may be expected that the latter will eventually influence the significance of the production centre positively. The advantage of the centre on the business is primarily caused by yield, quality or earliness. These are determined by the quality of the decisions at the operational level. It is not surprising that the interest of the growers with regard to the application of information technology in the glasshouse sector, and especially in the vegetable growing sector, is primarily directed at recording of data of the growing process and comparing these figures with their colleagues. In other parts of horticulture such developments are less spectacular, but they grow gradually. The answer to the question how far information technology will influence the function of the centre depends on the belief one has in a complete blueprint of the growing process. The information technology permits growers to make better observations on their farms and to complete the data faster and more adequately. The need to contact like minded colleagues to compare and discuss data and to exchange opinions for the sale of "fine-tuning" of the process, will increase.

In international competition, Dutch horticulture has sought strength in position-improving and less in competition-reducing strategies. The advantage of self-sustaining horticultural business groupings are evident. Willingness to cooperate between producers is thereby essential. The social-psychological aspects are now becoming more important. It is hopeful that the young generation is sufficiently equipped to do this. Abellas distinguishes in clustering his population of growers, a young and ambitious group of entrepreneurs.[9] He describes their management design as young, reflecting good education, strongly externally directed and exploiting a relatively large and modern business. There are also advantages in the centres as to the pressure of the environment. On the whole it is to be expected that future export-oriented horticulture can only flourish in self-sustaining horticultural business groupings.

---

6 The Dutch Agricultural Research Institute 1984.

7 Agrarian Entrepreneurship in Psychological Perspective. Defares P.B. The Sutch Agricultural University W

8 Management in Horticulture under glass. Abellas J.T.W. The Dutch Agricultrual Economics Institute, 1984.

9 Analysis of Management of Horticulture under glass. Abellas J.T.W. The Dutch Agricultural Economics Institute. 1988

CHAPTER 7

THE MEDITERRANEAN EXPORTERS SPAIN, ITALY, GREECE AND PORTUGAL[1]

SPAIN

The supply of Spanish produce is of great importance to other countries in the EEC. Indeed Spain with Holland stand out as being the largest exporters of fruit and vegetables within the Community, Spain has emphasis on fruit, and in particular citrus fruit, grown in the Mediterranean climate, and Holland has emphasis on vegetables, particularly salad vegetables, much of which are grown under protection in a north temperate climate.[2 3 4]

Vegetable Production in Spain

The Spanish area and production of vegetables are given in Table 1. Spain has a large area of land under vegetables, nearly half a million ha, with a total vegetable production of a little over ten million tonnes. Like the pattern of vegetable production in Italy the share of brassicas in the total vegetable area is low, some seven per cent, of which is cauliflowers are practically one half. Vegetables, leafed or stalked, account for a 19 per cent share of the Spanish area of vegetables. Of the 91 thousand ha of leafy and stalked vegetable crops which cover over 20 thousand ha are lettuce, asparagus and artichokes. As in Italy, the largest share of the Spanish area of vegetables is that cultivated for fruit, 43 per cent. The largest area, 67 thousand ha is under melons, with tomatoes second at 56 thousand ha, much of which are processed into tomato paste. Water melons and peppers are equal, at 27 thousand ha each. Some 18 per cent of the Spanish vegetable area is in roots and tubers, of which the garlic crop is one half or 42 thousand ha. This is followed by onions at 31 thousand ha which is a large export crop. The climatic conditions and the value in export market of garlic and

---

1 In strict terms of geography, Portugal, of course, is not a Mediterranean country, but in climate and in the pattern of production of fruit and vegetables and their trade, Portugal is a Mediterranean country like Spain, Italy and Greece. Perhaps it would be more correct to cite Portugal as being like the other countries in the Mediterranean zone, Spain, Italy and Greece.

2 EEC Enlargement in the Fruit and Vegetable Sector. Hinton Lynn. Price and Market Policies in European Agriculture Ed. Thompson K.G. and Warren R.M. Department of Agricultural Economics. University of Newcastle upon Tyne pp234-247

3 Spanish Accession to the EEC. Prospects for Temperate Fruit and Vegetables. Acta Horticulturae no. 955 pp447 - 414, 1984

4 EEC Enlargement in the Fruit and Vegetable Sector, with special particular reference to Spain and the UK. Hinton, Lynn Occasional Paper no. 31, Department of Land Economy, 1983

Table 1. Spain - Vegetable Production, Areas and Quantities, 1987

Items as listed in Eurostat

	'000 ha	'000 tonnes
Cauliflowers	11.4	234.2
Brussels sprouts	0.6	11.8
Cabbage	8.0	221.3
Other brassicas	12.6	282.8
All brassicas	**32.6**	**750.1**
Celeriac and celery	2.1	55.1
Leeks	3.0	61.9
Lettuce	26.1	624.4
Endive	3.7	77.7
Spinach	3.4	54.7
Asparagus	21.9	82.3
Chicory	0.2	3.4
Globe artichoke	25.3	338.0
Other vegetables (leafed or stalked)	5.4	115.4
Vegetables leafed or stalked	**91.1**	**1412.9**
Tomatoes	56.1	2447.3
Cucumbers	6.1	333.6
Gherkins	1.9	20.8
Melons	66.6	908.6
Water melons	27.4	627.4
Aubergines	4.4	126.0
Pumpkins, marrows	7.1	229.9
Peppers	27.7	705.9
Cultivated for fruit	**197.3**	**5399.5**
Turnips	1.5	23.6
Carrots	6.2	179.2
Garlic	42.5	238.4
Onions	31.0	1110.1
Beetroot	0.6	13.7
Radish	1.0	14.1
Other roots and tubers	2.1	32.8
Roots and tubers	**84.9**	**1611.9**
Peas	10.3	22.8
French beans and runner beans	26.3	264.7
Other pulses	18.1	159.8
Pulses	**54.7**	**447.3**
Cultivated mushrooms	**0.2**	**48.0**
Other fresh vegetables	**7.6**	**105.5**
Total of vegetables specified	**468.4**	**9775.2**

Source: Eurostat 1989

Table 2. Spain - Fruit (excluding citrus) Production, Areas and Quantities, 1987

Items as listed in Eurostat

	'000 ha	'000 tonnes
Fresh grapes	**68.0**	**520.1**
Apples	60.4	1042.9
Pears	33.9	520.6
Apples and pears	**94.3**	**1563.5**
Peaches	61.9	545.2
Apricots	22.1	141.8
Cherries	22.0	68.7
Plums	19.5	133.9
Nectarines	6.0	58.8
Stone fruit	**131.5**	**948.4**
Strawberries	10.7	231.3
Kiwi	0.8	0.7
Other soft fruit	10.0	74.9
Soft fruit	**21.5**	**306.9**
Oranges	137.3	2437.7
Mandarins	3.4	42.2
Satsumas	15.9	481.7
Clementines	43.9	769.5
Lemons	53.3	738.0
Grapefruit	0.9	18.2
Other citrus fruit	2.4	16.4
Citrus fruit	**257.1**	**4503.7**
Figs	20.2	55.7
Quinces	0.6	10.9
Avocados	5.9	32.5
Other fruit NOS	14.5	486.1
Other fruit	**41.2**	**585.2**
Total of fruit specified	**613.6**	**8427.8**

Source: Eurostat 1989

onions result in 87 per cent of the area of roots and tubers in Spain being given over to these two crops. Pulse crops represent 11 per cent of the area of vegetables with 26 thousand ha or almost half in French beans and runner beans with ten thousand ha in fresh peas. Of the ten million tonnes of vegetables in Spain, 5.7 million tonnes are vegetables grown for their fruit of which by far the largest share in tomatoes. Melons and water melons come some way behind and there is a large quantity of 706 tonnes of peppers, large for the nature of the commodity.

Fruit Production in Spain

The Spanish area and production of fruit are given in Table 2. Fruit production in Spain covers 0.6 million ha, virtually identical to that of Italy, but production by volume would appear to be less in Spain, however, there was a poor yield of grapes in Spain in the year shown.

Some similarity between the pattern of fruit production in Spain and that in Italy is to be expected. In Spain apples and pears are 15 per cent of the fruit area, as opposed to 22 per cent in Italy but pears are 35 per cent of the total apple and pear area in each country. Stone fruit is 21.5 per cent of the fruit area in Spain whereas it is higher in Italy at 27 per cent. Among stone fruit, peaches are the most important in both countries, approximately half the area of stone fruit. The area under apricots is larger in Spain but Spain also has a much smaller area of nectarines. Plums and cherries, both fruits which grow well in a north temperate climate, are prominent. However these crops grow better in a Mediterranean climate. Cherries and plums from the Mediterranean are on the market sooner than those grown in the north temperate climate zone. Soft fruit is much less prominent in Spain which grows fewer types of soft fruit, but strawberries grown in Spain are harvested at least a month before those in the north temperate zone, and Mediterranean strawberries are of high quality. Strawberries are prominent among the soft fruit in Spain and Italy. Citrus fruit in Spain, at 43 per cent of the total fruit area is of great importance in the fruit cropping. Citrus is 30 per cent of the fruit cropping in Italy. Citrus fruit in Spain, at 257 thousand ha, cover a 40 per cent larger area than in Italy. Oranges dominate the citrus fruit sector in both Italy and Spain, though the area in lemons is a larger in Italy. Grapefruit is quite a minor citrus fruit in both countries. While figs are grown in Spain and Italy, there is more in Spain.

Spanish Exports of Vegetables

Spanish exports of vegetables by country of destination for 1987 are illustrated in Figure 1 and the data is given in detail in Table 3. Much of Spain's vegetable production is partly exotic to countries in the north European climatic zone, while much of it is of higher quality because of the more favourable climate in Spain. Spain's large export of vegetables in round figures is 1.2 million tonnes. The Figure discloses that no less than ninety per cent of Spanish exports of vegetables goes to the EEC. West Germany and France each take 27 per cent of Spanish vegetable exports. France takes a large amount for a country with a sizable vegetable production, but her consumption is high and her proximity to Spain guarantees the freshness of produce. Britain takes 21 per cent of the Spanish exports of vegetables and other countries in the EEC take 16 per cent between them. The remaining ten per cent of the vegetable exports represent extra-EEC trade. Spain is so strong in vegetable exports that she may be regarded as the 'southern market garden' of Europe. Spain is the counterpart in vegetable production to Holland, which is geared to the same market, one substantially supplied by protected cultivation and the other substantially supplied from production grown under natural conditions without heat. Dutch and Spanish vegetables, and in particular the high price salads, are complimentary in the European market. For example, the Spanish tomatoes are marketed in the winter and the Dutch tomatoes in the summer. Naturally grown produce from Spain phases in, in its season, at the time protected vegetables from Holland are phased out.

The minor outlets for Spanish vegetables are the Netherlands, which takes 9 per cent, a large amount for a small country with a population of only 14 million. Belgium-Luxembourg takes less than 3 per cent, Italy a little more than 2 per cent, while Greece is seen to import only garlic from Spain. The low or absent exports of vegetables from Spain to Italy and Greece arises on account of their common climate and Italy and Greece consume a big proportion of home grown vegetables. Ireland,

Figure 1. Spanish Exports of Vegetables by Country of Destination

Exports by Quantity - 1987

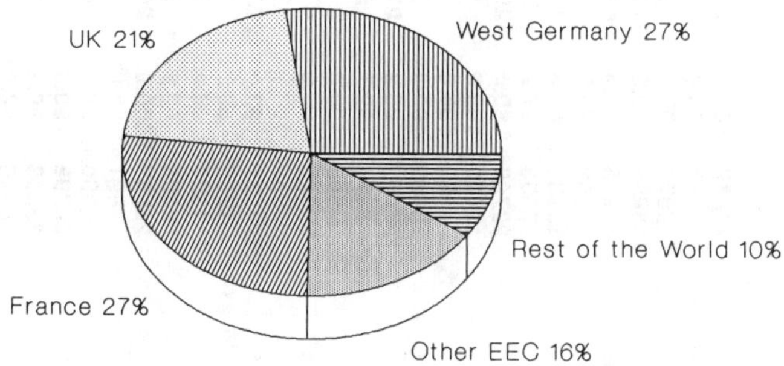

Source: Eurostat External Trade, Exports, 1989

Exports by Value - 1987

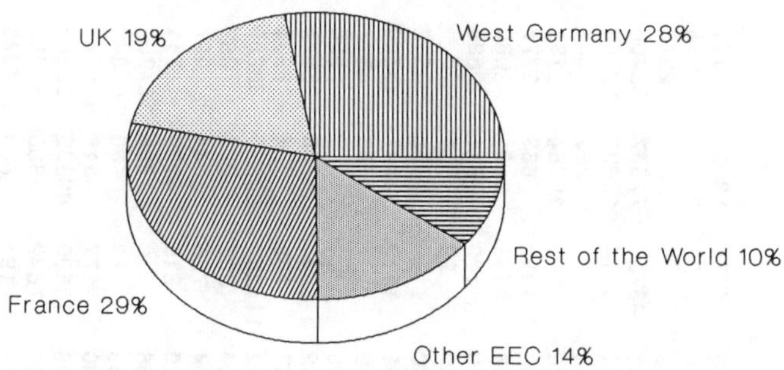

Source: Eurostat External Trade, Exports, 1989

Table 3. Spanish Exports of Vegetables by Country of Destination - 1987 tonnes

	TOTAL	UK	WG	FR	BL	NE	DE	IT	IR	GR	EEC	PER CENT	INTRA EEC	OTHER COUN- TRIES	PER CENT
Lettuce	84533	34774	18051	191	1447	14937	1461	1015	13		71889	85.0	71885	12644	15.0
Chicory	50	22		18		9		1			50	100.0	52		
Chard	2375			2375							2375	100.0	2413		
Tomatoes	247342	43736	77567	62408	1308	21833	3482	7865	236		218435	88.3	218437	28907	11.7
Cucumbers	62351	8714	10637	10852	815	15917	1260		50		48245	77.4	48384	14106	22.6
Gherkins	894		262	547							809	90.5	809	85	9.5
Other salad	22625	195	12464	131	1123	1107	1	214			15235	67.3	15239	7390	32.7
Cauliflowers	7549	1912	563	3777	33	963	88				7336	97.2	7371	213	2.8
Cabbages	25796	9563	3561	2069	115	3683	1769				20760	80.5	20778	5036	19.5
Brussels sprouts	1879	100	811	534		434					1879	100.0	1886		
Spinach	32	29	1			2					32	100.0	67		
Peas	3460		55	3319	28	58					3460	100.0	3751		
Beans	28891	1519	3747	15740	1342	2958		2502			27808	96.3	27885	1083	3.7
Other legumes	5389		90	5177	59	48					5374	99.7	5729	15	0.3
Celeriac	506			506							506	100.0	847		
Carrots	39562	6044	3768	26077	1612	139	786		694		39120	98.9	40076	442	1.1
Edible roots	3667	288	4	3335		24		16			3667	100.0	3667		
Onions	299152	114250	100375	30289	16838	13635	862	11	1911		278171	93.0	278301	20981	7.0
Garlic	3040	387	274	392	274	340		20		21	1708	56.2	1927	1332	43.8
Leeks	1354	725	44	237	14	302	10				1332	98.4	1332	22	1.6
Asparagus	11039	315	6279	711	655	135		1164			9259	83.9	9264	1780	16.1
Artichokes	26596		298	24752	94			1451			26595	100.0	26674	1	
Mushrooms	753	73	30	349	1			238			691	91.8	742	62	8.2
Fennel	6850	177	318	6248	10	77					6830	99.7	6849	20	0.3
Sweet peppers	224629	13590	88092	55885	3617	30305	2255	10357	197		204298	90.9	204345	20331	9.1
Courgettes	65730	7542	2660	47455	939	2512	72	2023			63203	96.2	63214	2527	3.8
Aubergines	8511	191	620	7040	155	383	31				8420	98.9	8443	91	1.1
Other vegetables	31129	10157	486	13034	728	3358	219	1629			29611	95.1	29610	1518	4.9
TOTAL	1215684	254303	331057	323448	31207	113159	12296	28506	3101	21	1097098	90.2	1099977	118586	9.8
Percent specified country of total		20.9	27.2	26.6	2.6	9.3	1.0	2.3	0.3	0.0	90.2			9.8	

Source: Eurostat External Trade, Exports, 1989

Figure 2. Spanish Exports of Fruit (excluding citrus) by Country of Destination

Exports by Quantity - 1987

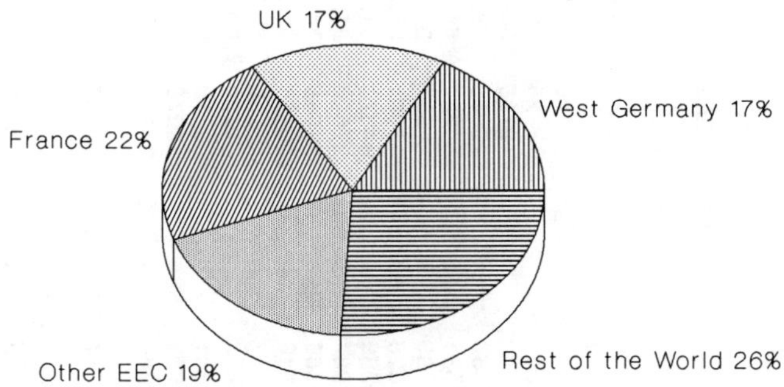

Source: Eurostat External Trade, Exports, 1989

Exports by Value - 1987

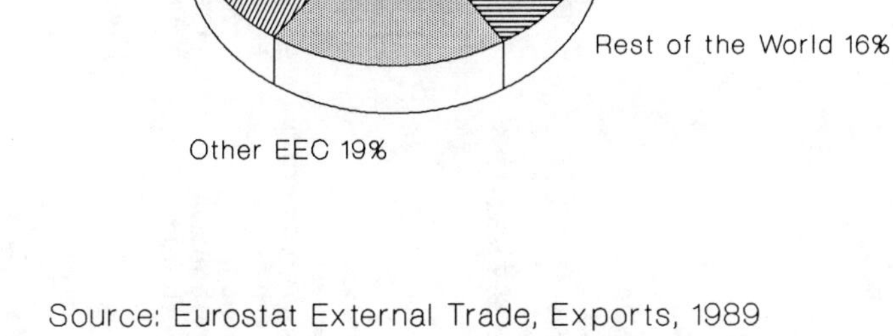

Source: Eurostat External Trade, Exports, 1989

Table 4. Spanish Exports of Fruit (excluding citrus) by Country of Destination, 1987, tonnes

	TOTAL	UK	WG	FR	BL	NE	DE	IT	IR	PO	EEC	PER CENT	INTRA EEC	OTHER COUN-TRIES	PER CENT
Bananas	2228	15				11					26	1.2	26	2202	98.8
Dates	71	67	3					1			71	100.0	4290		
Pineapples	120		120								120	100.0	121		
Kiwi fruit	24		24								24	100.0	59		
Avocados	10735		1122	9151	394	68					10735	100.0	10923		
Apricots	31679	955	2152	8299	1910	295		11458			25069	79.1	25090	6610	20.9
Peaches and nectarines	58578	5435	7075	27401	2723	1483	164	5891	115		50287	85.8	50491	8291	14.2
Cherries	929	66	30	614	47	8		86			851	91.6	1123	78	8.4
Apples	27555	4460	2433	946	1591	609	236	325	53	348	11001	39.9	11143	16554	60.1
Pears	139177	4000	11499	6694	519	1291	18	20352			44373	31.9	44378	94804	68.1
Plums	30433	8469	2367	6434	2614	4731	8	3067			27690	91.0	28358	2743	9.0
Table grapes	109286	25576	18185	19471	1258	6003	1451	3866	1858		77668	71.1	77981	31618	28.9
Strawberries	110111	9729	34611	38551	6591	8532	633	1518			100165	91.0	100496	9946	9.0
Raspberries,redcurrants	9			8							8	88.9	12	1	11.1
Other fresh berries	409		150	259							409	100.0	562		
Water melon	126999	9535	36625	41438	4830	5585	916	10341			109270	86.0	109595	17729	14.0
Other fresh melon	133280	62026	19815	9594	4247	17481	1233	998	685		116079	87.1	116553	17201	12.9
Other fresh fruit	17245	3426	352	4140	411	219		7944	4	615	17111	99.2	17118	134	0.8
TOTAL	798868	133759	136563	173000	27135	46316	4659	65847	2715	963	590957		598319	207911	
Percent specified countries of total		16.7	17.1	21.7	3.4	5.8	0.6	8.2	0.3	0.1	74.0		74.9	26.0	

Source: Eurostat External Trade, Exports, 1989

with the lowest per capita vegetable consumption in the EEC, is seen to take less than one per cent from Spain. Such a low figure suggests that Ireland receives some vegetables via Britain, not disclosed by the Eurostat export data. While not expecting Ireland to produce a lot of fruit and vegetables, she could be more self-sufficient in vegetables as well as fruit.

Countries other than those in the EEC are important in the Spanish exports of vegetables for a few commodities. Ten per cent or 14 thousand tonnes of cucumbers go to the profitable Scandinavian market, to Finland and to Switzerland and Austria. Thirteen thousand tonnes of lettuce goes to "other countries" of which Sweden takes over nine thousand tonnes. Of the 7400 tonnes of other salad vegetables exported 6740 tonnes go to Switzerland. Forty four per cent or 1332 tonnes of the export of garlic goes to "other countries", more than half to Czechoslovakia and a quarter to Finland. Garlic important in cooking to northern countries, is ideally grown in Spain. Of the five thousand tonnes of cabbages exported to 'other countries' Finland and Sweden each take 1500 tonnes. These countries are followed by Norway, Switzerland and Austria. 'Other countries' take 16.1 per cent or 1780 tonnes of Spanish exports of asparagus, most to Switzerland. Of exports of tomatoes to 'other countries', 29 thousand tonnes, most go to Scandinavia and Finland, but 55 thousand tonnes go to the United States in the summer period. Spanish exports of vegetables by value are also given in Figure 1. World exports of Spanish vegetables are 537 million ECUs and exports to the EEC are 466 million ECUs.

Spanish Exports of Fruit other than Citrus Fruit

Spanish exports of the above class of fruit are illustrated in Figure 2, both in volume and value terms, and given in detail in Table 4. Seventy six per cent in volume terms of the Spanish exports of fruit excluding citrus go to other countries in the EEC and France takes the biggest share of 22 per cent. This is partly on account of France being adjacent to Spain, while Spain has an advantage in climate and there is a strong demand in France for fruit. West Germany takes 17 per cent of Spanish fruit exports other than citrus and Britain, another 17 per cent. Minor markets in the EEC are Italy, eight per cent, the Netherlands, six per cent, Belgium-Luxembourg, 3.4 per cent, while Ireland and Portugal take less than one per cent. Spanish exports of non-citrus fruit amount to 0.8 million tonnes.

Twenty six per cent of fruit exports other than citrus go to other countries outside of the EEC. Three-quarters of the Spanish exports of apricots go to "other countries". Five thousand tonnes or 16 per cent go to Switzerland, with smaller quantities going to Austria and the Canary Islands. Sixty per cent of all Spanish exports of apples, 16 thousand tonnes, go to countries outside the EEC. Over half of these go to the Canary Islands, which unlike Spain is not a member of the EEC. There is however, a special relationship between Spain and the Canary Islands, where the temperature is too high for growing apples. Similarly of the sixty eight per cent of the exports which go to "other countries", most go to the Canary Islands and a small quantity go to Scandinavia, Finland and Austria. Almost one third of the Spanish exports of table grapes go to "other countries" than the EEC. Sweden and Norway each take seven per cent, while Finland, Austria, Switzerland and the Canary Islands each take three per cent. Canada, quite a remarkable outlet for Spain because of the distance involved, takes four per cent. The small Spanish exports of bananas go to Ceuta, which lies just over the straights of Gibraltar.

World exports of Spanish fruit other than citrus are 456 million ECUs and exports to the EEC are 380 ECUs or 83 per cent of all exports of non-citrus fruit.

Spanish Exports of Citrus Fruit

Citrus fruit exports from Spain are illustrated in Figure 3 and shown in detail by volume in Table 5. Eighty three per cent of the Spanish citrus exports go to countries in the EEC. France in 1987 took the most at 30.2 per cent, followed by West Germany with 29.9 per cent. The United Kingdom follows a long way behind the leading markets with 9.0 per cent. Holland is next with 7.1 per cent then Belgium-Luxembourg with 5.6 per cent. Denmark follows with 1.1 per cent, then Ireland with 0.3 per cent and Greece with only 772 tonnes of the Spanish export of lemons, too small to register one per cent. Italy does not import citrus fruit from Spain. The absence of Spanish citrus imports to Italy and the

Figure 3. Spanish Exports of Citrus Fruit by Country of Destination

Exports by Quantity - 1987

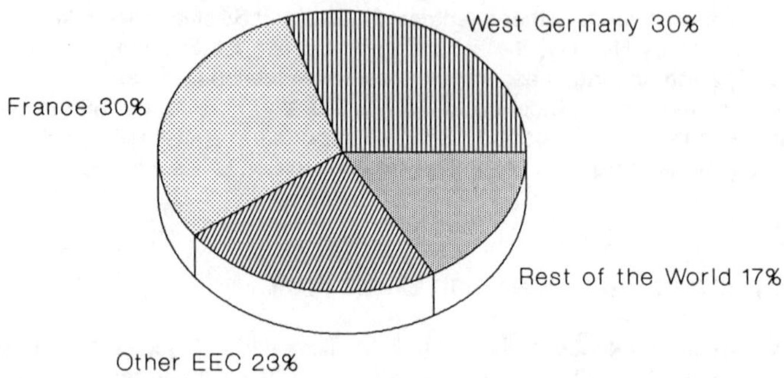

Source: Eurostat External Trade, Exports, 1989

Exports by Value - 1987

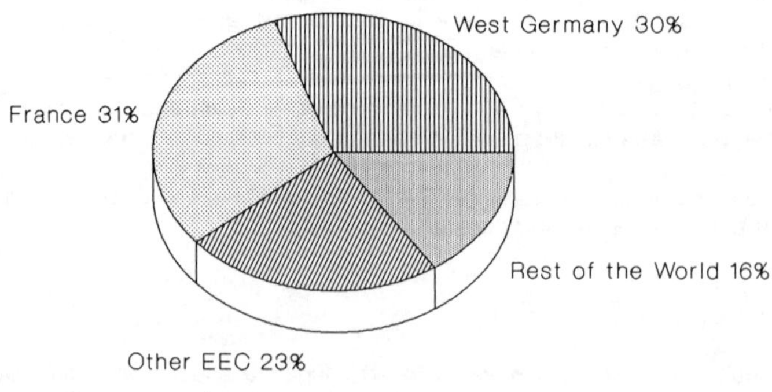

Source: Eurostat External Trade, Exports, 1989

Table 5. Spanish Exports of Citrus Fruit by Country of Destination - 1987 tonnes

	TOTAL	UK	WG	FR	BL	NE	DE	IR	GR	EEC	PER CENT	INTRA EEC	OTHER COUN-TRIES	PER CENT
Oranges	1154787	73672	348633	369155	80371	105077	9784	1165		987857	85.5	988247	166930	14.5
Other sweet oranges	797	546	85	13	111			20		775	97.2	513	22	2.8
Other than sweet	3838	3838								3838	100.0	4583		
Sanguines	8152	28	89	7912	62	55				8146	99.9	8888	6	0.1
Clementines	567412	26240	191260	217021	24508	36745	12261	436		508471	89.6	508576	58941	10.4
Monreales and Satsumas	241219	78949	70418	4599	12803	12093	921	4412		184195	76.4	184194	57024	23.6
Mandarins and Wilkins	2870	531	298	1830	211					2870	100.0	2956		
Other citrus hybrids	12327	2140	5152	2159	950	558				10959	88.9	11009	1368	11.1
Lemons	414569	30137	106315	121290	16178	16137	4086	229	772	295144	71.2	295158	119425	28.8
Grapefruit	7150	1116	222	5270	390	150	2			7150	100.0	7150		
TOTAL	2413121	217197	722472	729249	135584	170815	27054	6262	772	2009405	83.3		403716	16.7
Percent specified countries of total		9.0	29.9	30.2	5.6	7.1	1.1	0.3	0.0	83.3			16.7	

Source: Eurostat External Trade, Exports, 1989

negligible amount to Greece, is to be expected because of their similar climates and production. However, the export of citrus fruit to the United Kingdom is notably low, in contrast with the larger export market in France and West Germany. The Netherlands has a population one quarter that of Britain, which means that the Dutch consume a great deal more citrus than the British. All the oranges other than sweet oranges go to the EEC, as also all the mandarins and satsumas and all the grapefruit. Of the 1154 thousand tonnes of oranges exported from Spain, 348.6 thousand tonnes go to West Germany and 369.2 thousand tonnes go to France. Belgium-Luxembourg in addition to the Netherlands takes more oranges than Britain. France takes by far the most of the tangerines and a third of the large export of 567.4 thousand tonnes of clementines, while Britain takes as much as a third of the export of mandarins, monreales and satsumas. Grapefruit with total exports of 7150 tonnes, is most popular in France, which takes nearly three quarters of the total, and the next most popular is Britain which takes over a thousand tonnes. Seventeen per cent of Spanish citrus goes to countries other than those of the EEC. For instance, fourteen per cent of Spanish orange exports of 170 thousand tonnes go to 'other countries'. Thus the Canary Islands take 4.3 per cent and Austria and the German Democratic Republic each take between one and three per cent, while Switzerland and Sweden each take 1.3 per cent.

A number of countries take less than one per cent of the orange exports. These are, Norway, 0.3 per cent, Finland, 0.1 per cent, Poland, 0.2 per cent, and Hungary with only 94 tonnes. Six tonnes of sanguines go to Switzerland, as do all the exports to third countries of 'other oranges'. Of the 59 thousand tonnes of clementines exported to third countries the most go to Switzerland, and smaller amounts go to the Canary Islands, Austria, Finland, Sweden, Norway, the United States and Canada. Of the 57 thousand tonnes of monreales and satsumas exported to third countries, most, 5.3 per cent of all exports go to Austria, 4.9 per cent go to Sweden, 4.1 per cent go to Czechoslovakia, 3.8 per cent go to Switzerland and another 3.8 per cent go to Finland. Some one thousand tonnes of "other citrus" go to Finland and Sweden. A big share of Spain's lemon exports go to third countries, so that the lemon export is also shared by a number of countries. The Canary Islands, Switzerland, Austria, Czechoslovakia and the German Democratic Republic each take some four per cent. Of the rest the Soviet Union takes 2.9 per cent, Canada and the United States, take smaller amounts.

Exports of citrus fruit by value, shows that the value of citrus fruit exported from Spain in 1987 was 971 million ECUs, 149 million to third countries and 816 million ECUs or eighty per cent to EEC countries.

ITALY

The Italian Production of Vegetables

Vegetable production in Italy, by area and production, is shown in Table 6. Italy grows a large area and a large range of vegetables covering over half a million ha. Italy's production of vegetables is correspondingly large at 12 million tonnes. Over nine per cent of the vegetable area is in brassicas, of which cauliflower is the most prominent, grown in the north of Italy. Over a quarter of the total vegetable area in Italy is down to leafed or stalked vegetables, of which lettuce, chicory and endive are the most important. Somewhat over forty per cent of the vegetables in Italy are vegetables cultivated for their fruit. This produce is mainly of high value and its large share is consistent with a Mediterranean climate. Partly as a consequence the share in the total vegetable area of roots and tubers is only seven per cent. Pulses amount to a fifteen per cent share and fresh peas, French beans and runner beans together cover three quarters of the total area of pulse crops. Many of Italy's vegetables are harvested early in the season and sold at good prices, like those of Spain, Greece and Portugal.

Italian Exports of Vegetables

The pattern of Italian vegetable exports is shown in Figure 4 and in detail in Table 7. Italy exports nearly 700 thousand tonnes of vegetables, somewhat less than half of her exports of non-citrus fruit. Similar to Italian non-citrus fruit exports, 85 per cent of Italy's vegetable exports go to other countries in the EEC. West Germany alone takes 47 per cent of the Italian vegetable exports. France takes 23.4 per cent of the Italian's vegetable exports, or twice the proportion of the French import of non-citrus fruit in which France is relatively well endowed. Of the other member states the UK takes nearly five per cent and the Netherlands takes a similar amount. Belgium-Luxembourg takes 3.3 per cent, Denmark 1.6 per cent and Ireland 0.5 per cent.

A fairly wide range of countries receive the 15 per cent of vegetable exports marketed outside the Community. The Canary Islands, which as mentioned, has a special relationship with Spain, are only seen to take one consignment of 4 tonnes of onions from Italy. Owing to its proximity to Italy, Switzerland with its affluent population and very low self-sufficiency in vegetables, takes 14.4 per cent. Switzerland takes almost the full range of Italian vegetable exports, which include over seven thousand tonnes of cauliflowers, some six thousand tonnes of salad vegetables, and four thousand tonnes of sweet peppers. Switzerland also depends on Italy for onions, gherkins and tomatoes and takes half Italy's export of mushrooms. Sweden takes a quarter of the Italian export of leeks and Austria takes 15 per cent of the Italian export of garlic. Czechoslovakia is seen to take 19 tonnes of garlic and nothing else. This is a comment on Czechoslovakian trade policy for fruit and vegetables, as no doubt, Czechoslovakia only imports the commodities she must.

Some 1300 tonnes of salad vegetables go to the USA, together with 27 tonnes of sweet peppers and very small consignments of lettuce and chicory. Smaller amounts of chicory and cauliflower go to Canada, while Japan also receives a small consignment of cauliflower and sweet peppers.

The Italian Production of Fruit

Fruit production in Italy, by area and production, is shown in Table 8. Thirty per cent of the total area of fruit in Italy is accounted for by citrus fruit. Some 428 thousand ha are in non-citrus fruit which has a production of 7.2 million tonnes. Apples and pears rank at just over one fifth of the fruit area and one third of the production of fruit by volume. Pears are prominent in Italy's fruit production making up 35 per cent by area of the small group of apples and pears. This proportion of pears is much higher than in northern countries of the EEC and the pears comprise a significant proportion of summer pears which are harvested early in the season.

Stone fruit accounts for somewhat over one fifth of the total fruit area, and one fifth of the total fruit production by volume in Italy. Again this is a much higher proportion than in northern countries.

Table 6. Italy - Vegetable Production, Areas and Quantities, 1987

Items as listed in Eurostat	'000 ha	'000 tonnes
Cauliflowers	31.7	598.1
Brussels sprouts	0.4	7.5
Cabbage	5.4	114.8
Other brassicas	9.2	209.4
All brassicas	**46.7**	**929.8**
Celeriac and celery	4.7	154.3
Leeks	1.3	36.5
Lettuce	20.0	386.3
Endive	13.6	262.1
Spinach	7.7	96.8
Asparagus	4.9	24.1
Chicory	14.1	208.6
Globe artichoke	48.2	453.2
Other vegetables (leafed or stalked)	22.4	515.6
Vegetables leafed or stalked	**136.9**	**2137.5**
Tomatoes	112.3	4860.0
Cucumbers	3.5	93.1
Gherkins	1.5	23.7
Melons	17.1	356.8
Water melons	21.0	714.7
Aubergines	11.2	295.6
Pumpkins, marrows	12.7	342.9
Peppers	16.1	387.7
Cultivated for fruit	**195.4**	**7074.5**
Turnips	2.8	45.6
Carrots	9.3	327.3
Garlic	5.0	46.2
Onions	18.5	478.1
Beetroot	1.3	30.6
Radish	2.2	50.0
Roots and tubers	**39.1**	**977.8**
Peas	30.0	64.4
French beans and runner beans	31.0	181.7
Other pulses	18.8	188.3
Pulses	**79.8**	**434.4**
Cultivated mushrooms	**-**	**73.3**
Other fresh vegetables	**0.6**	**14.0**
Total of vegetables specified	**498.5**	**11641.3**

Source: Eurostat 1989

Figure 4. Italian Exports of Vegetables by Country of Destination

Exports by Quantity - 1987

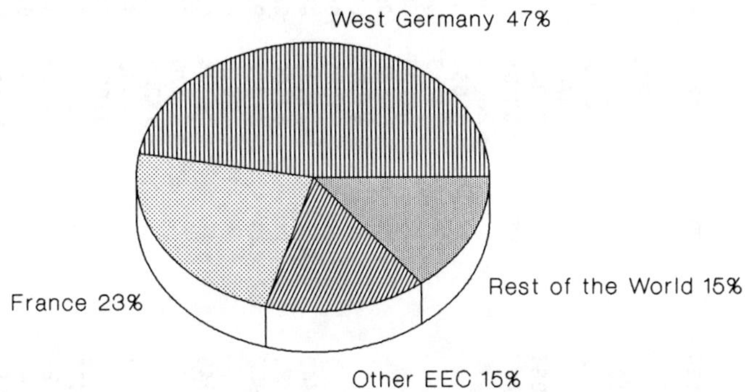

Source: Eurostat External Trade, Exports, 1989

Exports by Value - 1987

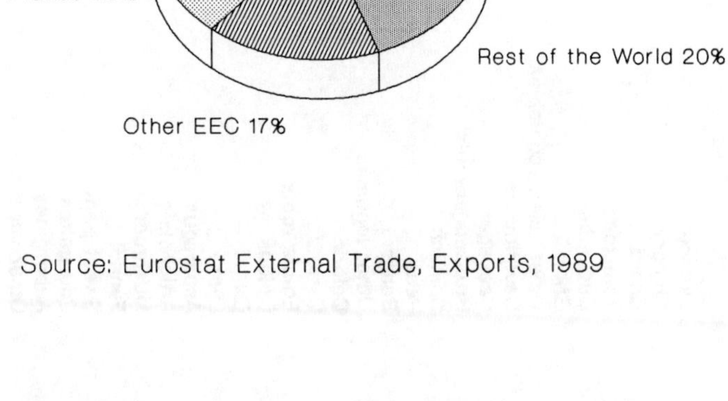

Source: Eurostat External Trade, Exports, 1989

Table 7. Italian Exports of Vegetables by Country of Destination - 1987 tonnes

	TOTAL	UK	WG	FR	BL	NE	DE	SP	IR	GR	EEC	PER CENT	INTRA EEC	OTHER COUN-TRIES	PER CENT
Lettuce	14244	1332	7261	2107	143	716	136		93		11788	82.8	11800	2456	17.2
Chicory	819		67	447	7						521	63.6	521	298	36.4
Chard	242			242							242	100.0	424		
Tomatoes	20917	56	13928	3984	474	31	129				18602	88.9	18601	2315	11.1
Cucumbers	5044	3	3947	278	3	13	2				4246	84.2	4244	798	15.8
Gherkins	7134		4537	219	4						4760	66.7	4794	2374	33.3
Other salad vegetables	97536	2136	43325	21667	3833	6735	251				77947	79.9	78121	19589	20.1
Cauliflowers	86821	2904	39347	10591	3026	10480	1733		53		68134	78.5	68135	18687	21.5
Cabbages	26683	4244	14075	2422	565	2076	336				23718	88.9	23931	2965	11.1
Brussels sprouts	711	60	368	40	12	101					581	81.7	612	130	18.3
Spinach	4745	476	3885		253	131					4745	100.0	5234		
Peas	1268		328	708	200	32					1268	100.0	2280		
Beans	2636	34	346	1992	73	60					2505	95.0	2536	131	5.0
Other legumes	12536		5016	2092	806	3116					11030	88.0	11234	1506	12.0
Celeriac	1391		642	623	126						1391	100.0	2015		
Carrots	115969	8787	51358	41965	2044	166	4443		2420		111183	95.9	111228	4786	4.1
Edible roots	32286	1525	25794	2905	371	1475	30				32100	99.4	32100	186	0.6
Onions	57959	1717	28247	17020	1426	1949	1989	2	128		52478	90.5	52565	5481	9.5
Garlic	3472	595	1406	365	130	316				16	2828	81.5	2850	644	18.5
Leeks	11970	178	4910	659	92	240	345				6424	53.7	6427	5546	46.3
Asparagus	306	41	93	42	2						178	58.2	178	128	41.8
Artichokes	14728		642	12823	740						14205	96.4	14405	523	3.6
Mushrooms	1217	13	301	203	17	3	3				540	44.4	389	677	55.6
Fennel	24047	783	5743	9064	1584	683					17857	74.3	17924	6190	25.7
Sweet peppers	38960	807	27102	2356	1076	972	363		2		32678	83.9	32706	6282	16.1
Courgettes	10254	1272	3178	3827	488	73	28				8866	86.5	8870	1388	13.5
Aubergines	12111	70	4313	5658	421	96	41				10599	87.5	10600	1512	12.5
Other vegetables	78119	6185	32360	15922	4514	2230	818		521		62550	80.1	62592	15569	19.9
TOTAL	684125	33218	322519	160221	22430	31694	10647	2	3217	16	583964	85.4	587316	100161	14.6
Percent specified country of total		4.9	47.1	23.4	3.3	4.6	1.6	0.0	0.5	0.0	85.4		85.8	14.6	

Source: Eurostat External Trade, Exports, 1989

Table 8. Italy - Fruit Production, Areas and Quantities, 1987

Items as listed in Eurostat

	'000 ha	'000 tonnes
Fresh grapes	**79.3**	**1588.8**
Apples	88.1	2235.9
Pears	48.3	894.4
Apples and pears	**136.4**	**3130.3**
Peaches	80.0	1190.2
Apricots	15.6	191.9
Cherries	25.3	144.5
Plums	12.3	147.8
Nectarines	28.4	323.4
Other stone fruit	0.7	7.3
Stone fruit	**162.3**	**2005.1**
Black and red currants	0.1	0.5
Strawberries	11.0	191.4
Raspberries	0.3	1.8
Kiwi	13.4	115.3
Other soft fruit	11.0	112.6
Soft fruit	**35.8**	**421.6**
Oranges	107.0	1342.9
Mandarins	15.9	191.4
Clementines	17.5	142.5
Lemons	39.7	591.8
Grapefruit	0.3	3.1
Other citrus fruit	2.9	42.1
Citrus fruit	**183.3**	**2313.8**
Figs	13.3	42.8
Quinces	0.1	1.0
Other fruit NOS	0.8	5.1
Other fruit	**14.2**	**48.9**
Total of fruit specified	**611.3**	**9508.5**

Source: Eurostat 1989

Figure 5. Italian Exports of Fruit (excluding citrus) by Country of Destination

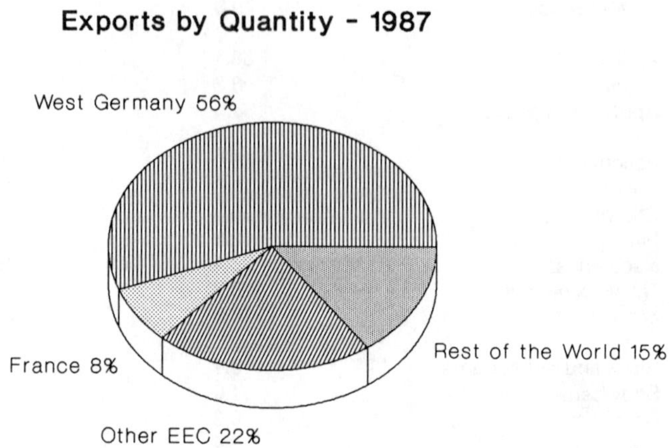

Source: Eurostat External Trade, Exports, 1989

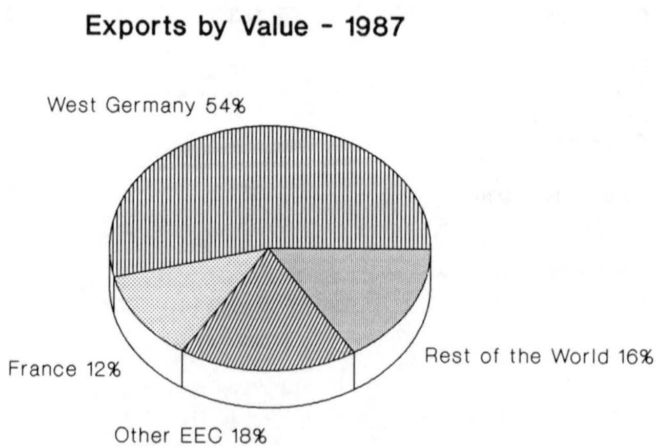

Source: Eurostat External Trade, Exports, 1989

Table 9. Italian Exports of Fruit (excluding citrus) by Country of Destination, 1987, tonnes

	TOTAL	UK	WG	FR	BL	NE	DE	SP	IR	PO	EEC	PER CENT	INTRA EEC	OTHER COUNTRIES	PER CENT
Bananas	119	38	9	32	21						100	84.0	100	19	16.0
Dates	168	2	123	35							160	95.2	161	8	4.8
Pineapples	606		58	48				500			606	100.0	658		
Kiwifruit	12439	448	6498	1641	859	330	281	307	93		10457	84.1	10556	1982	15.9
Avocados	37			37							37	100.0	37		
Mangoes, guavas and mangosteens	2429		1086	192	78						1356	55.8	1544	1073	44.2
Apricots	15589	396	5706	1203	459	308					8072	51.8	8150	7517	48.2
Peaches and nectarines	433365	71403	215995	17917	27884	16229	9367	733	3075		362603	83.7	362660	70762	16.3
Cherries	14538	1483	9476	74	941	622	284				12880	88.6	12893	1658	11.4
Other stone fruit	14477		12518			372					12890	89.0	14858	1587	11.0
Apples	328267	8480	261011	13735	2743	5143	9626	389	384	150	301661	91.9	301689	26606	8.1
Pears	70496	4878	38912	7803	31	306	2995	2252	1815		58992	83.7	59331	11504	16.3
Plums	9006	1878	4300	703	463	501	274				8119	90.2	8232	887	9.8
Table grapes	470346	22897	208340	112904	28884	9583	6644		316		389568	82.8	389747	80778	17.2
Strawberries	55787	730	41330	816	385	1159	1033				45453	81.5	45485	10334	18.5
Raspberries,redcurrants	162		130	12							142	87.7	141	20	12.3
Blackcurrants	20		20								20	100.0	20		
Cranberries	26												14	26	100.0
Other fresh berries	155		59	96							155	100.0	170		
Water melon	65012	2137	30958	17291	1692	2463	1309				55850	85.9	55971	9162	14.1
Other fresh melon	10748	418	2668	1363	961	1601	487		3		7501	69.8	7550	3247	30.2
Other fresh fruit	5140	185	1776	1199	481	91			28		3760	73.2	3934	1380	26.8
TOTAL	1508932	115373	840973	177101	65882	38708	32300	4181	5714	150	1280382	84.9	1283901	228550	15.1
Percent specified countries of total		7.6	55.7	11.7	4.4	2.6	2.1	0.3	0.4	0.0					

Source: Eurostat External Trade, Exports, 1989

It includes a large area of peaches, cherries and nectarines suited to the Mediterranean climate. These fruits, some of which are also grown in the northern countries of the EEC grow better and are more early when grown in a Mediterranean climate.

Because the pattern of fruit production is so much determined by the Mediterranean climate the area of soft fruit is only 6 per cent of the Italian fruit area. Soft fruit accounts for between four and five per cent of the fruit production by volume. Fresh grapes cover only 1.3 per cent of the fruit area, but 17 per cent of the fruit production by volume. There is a small production of figs.

Citrus fruit is ideally produced in a Mediterranean climate with little risk of frost. Nearly one quarter of Italy's entire fruit production by volume is citrus fruit, of which the orange is more than half, but mandarins and clementine are also prominent. Lemons are next to oranges, with a quarter of the citrus fruit by weight and 21 per cent by area. Grapefruit, are seen to be small in area and production as compared with their frequency of availability on the market stall or in the supermarket, in the UK.

Exports of Fruit other than Citrus Fruit

Figure 5 shows that while 85 per cent of the Italian non-citrus fruit exports go to other member states of the EEC, 56 per cent goes to West Germany alone. From the data on exports of non-citrus fruit given in Table 9, of total exports of 1.5 million tonnes, practically 12 per cent goes to France, between seven and eight per cent go to the UK, and 4.4 per cent go to Belgium-Luxembourg. The Netherlands take only 2.6 per cent, Denmark takes 2.1 per cent and Ireland, Spain and Portugal take less than one per cent between them. These latter countries and Greece, for which no exports from Italy are recorded, are all in the Mediterranean zone and so are in direct competition with Italy.

The 15 per cent of fruit excluding citrus fruit, going outside the EEC, is sent to a great range of countries, a much greater range than those of Spanish exports. Big buyers are Austria and Switzerland, which, of course, have a common border with Italy. This means low transport costs and the produce gets to these markets fresh. Austria takes virtually 70 thousand tonnes of fruit, excluding citrus, and Switzerland 86 thousand tonnes, together 156 thousand tonnes, of the 229 thousand tonnes exported to countries other than those in the EEC. Norway takes nearly 14 thousand tonnes, Finland takes over nine thousand tonnes, and Saudi Arabia takes nearly as much. Countries which take less than one thousand tonnes are numerous. These include the Canaries, Czechoslovakia, the United Arab Emirates and Malta. Canada takes 1260 tonnes of table grapes, together with 29 tonnes of pears, and Czechoslovakia takes only a token of 79 tonnes of peaches. In general, few consignments of luxury fruit and vegetables from Spain and Italy go to countries in the Eastern block. This suggests that a small elite with purchasing power is receiving these luxury fruit and vegetables in the Eastern block. It will be seen that Greece exports quite a significant amount of fruit and vegetables to East Europe.

Looking at the relative importance between commodities in the exports of non-citrus fruit, table grapes stand out at 47 thousand tonnes. They are followed by peaches and nectarines at 43 thousand and apples at 33 thousand tonnes. These commodities alone are an 80 per cent share of the exports of Italian non-citrus fruit.

Italian Exports of Citrus Fruit

The pattern of the exports of citrus fruit is revealed in Figure 6 and in Table 10. This shows Italy as an even more international supplier for citrus fruit than she is for non-citrus fruit. Over half of Italy's export of citrus fruit goes to countries outside the EEC. West Germany takes the biggest of the EEC share, but only 31 per cent. The traditional ties between the UK and Spain and between Italy and West Germany in the fruit and vegetable trade have been mentioned. It would seem that the trading bond between West Germany and Italy is much weaker than that between the UK and Spain. Oranges of different kinds account for nearly 0.15 million tonnes out of a citrus export total of 0.2 million or nearly 70 per cent of the total. 81 per cent are of the blood orange type. The popularity of the blood orange is quite different from one country to another. In West Germany they are 64 per cent of the orange supply, while in Belgium-Luxembourg sanguines are 82 per cent of the orange supply. In Britain they are 65 per cent of the supply, and in France, 33 per cent. While West Germany is the

Figure 6. Italian Exports of Citrus Fruit by Country of Destination

Exports by Quantity - 1987

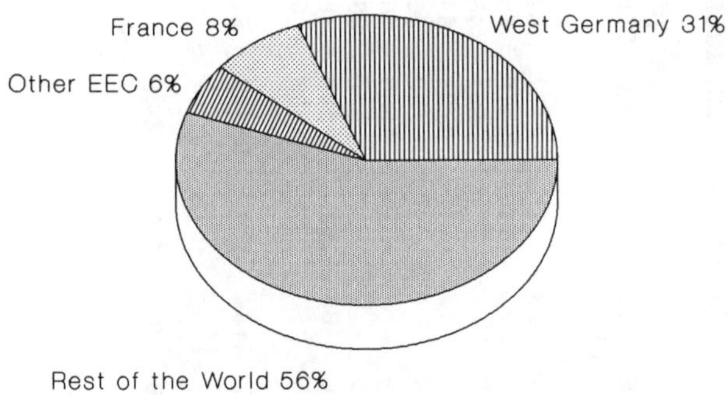

Source: Eurostat External Trade, Exports, 1989

Exports by Value - 1987

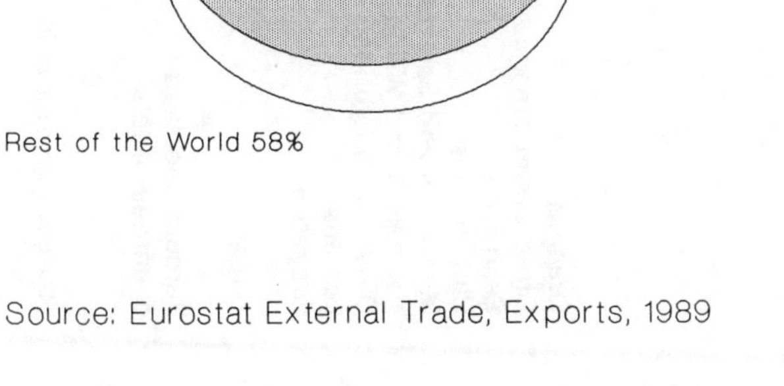

Source: Eurostat External Trade, Exports, 1989

Table 10. Italian Exports of Citrus Fruit by Country of Destination - 1987 tonnes

	TOTAL	UK	WG	FR	BL	NE	DE	IR	EEC	PER CENT	INTRA EEC	OTHER COUN-TRIES	PER CENT
Oranges	2312		1564	520	43	11	28	1	2167	93.7	2170	145	6.3
Other sweet oranges	22314	327	9812	8631	595			11	19376	86.8	21410	2938	13.2
Sanguines	117542	3040	42325	5685	4356	1455			56861	48.4	58229	60681	51.6
Clementines	1892	220	1474	43	16	19	18	1	1791	94.7	1791	101	5.3
Monreales and Satsumas	21		2			18			20	95.2	20	1	4.8
Mandarins and Wilkins	2832	541	71	496	177				1285	45.4	1840	1547	54.6
Other citrus hybrids	525	92	18	212		195			517	98.5	518	8	1.5
Lemons	64450	451	9605	1008	120	154	88	163	11589	18.0	11590	52861	82.0
Grapefruit	1193		730	454	4	5			1193	100.0	1193		
TOTAL	213081	4671	65601	17049	5311	1857	134	176	94799	44.5	98761	118282	55.5
Percent specified countries of total		2.2	30.8	8.0	2.5	0.9	0.1	0.1	44.5			55.5	

Source: Eurostat External Trade, Exports, 1989

biggest export market for citrus fruit, France is the next most important market in the EEC, taking eight per cent of the exports of Italian citrus fruit. The UK and Belgium take 2.2 and 2.5 per cent respectively, but other member states take considerably less. Reference to the pattern of citrus exports from Spain shows that the EEC member states are getting most of their citrus supplies from Spain. Lemon exports of 64 thousand tonnes are 30 per cent of all citrus exports from Italy. These are substantially sent to countries outside the EEC, over 80 per cent. No less than a third of this quantity goes to Poland, 11 per cent to Hungary and ten per cent go to Austria. Smaller amounts go to Switzerland, Yugoslavia and Romania. Fifty two per cent of the sanguines go to non-EEC countries. Nearly half go to Switzerland and nine per cent in each case go to Austria and Hungary. It is true that East European countries do get supplies of citrus fruit which of course they cannot grow, but the supply of their produce from Italy is very small. The East European market, it will be seen, is met more substantially by Greece. They are prepared to import citrus fruit from Italy but their imports of vegetables from Italy is practically non-existent.

GREECE

Vegetable Production in Greece

The area of fruit and vegetable production in Greece is much less than, say that in Spain because the area of cultivatable land in Greece is more limited. Thus the vegetable area in Greece is 29 per cent of that in Spain, and the area in fruit production is 36 per cent of that in Spain.

As with the other Mediterranean exporters there is a wide range of vegetables on offer, as shown in Table 11. Similar to other countries too, the area of brassicae in Greece is small, only eight per cent, with cauliflower in the minority to cabbage. Leafed and stalked vegetables comprise 12 per cent of the vegetable area in Greece. Vegetables listed under leafed and stalked are identical to the list for Spain, except for a small area of chicory, which is included under "other vegetables" for Greece. However, because of the much larger ha of leafed and stalked vegetables in Spain, lettuce, spinach and globe artichoke are some 20 thousand hectares to become between two and three thousand in Greece. The pattern of vegetable production is somewhat different than that in Spain. In particular in Greece 60 per cent or 83 thousand ha of the vegetable area is given to vegetables cultivated for their fruit, the most valuable sector of Mediterranean vegetable production. The proportion in Spain is 47.5 per cent. Tomatoes in Greece are more prominent in their group of vegetables than they are in Spain and melons feature little in Greece, compared to Spain. Roots and tubers account for nearly 12 per cent of the vegetable area, or 16 thousand ha, much less than the 85 thousand ha in Spain. In Greece, onions, by contrast with Spain, dominate the cropping in the root and tuber sector, 12 thousand ha of a total of 16 thousand ha, and the composition of the roots and vegetables is quite different from that in Spain. Pulse crops are also low in proportion in Greece, as compared with Spain, but French and runner beans are the most popular pulse crops in both countries. In terms of production by volume, Greece grows 3.7 million tonnes of vegetables as compared with Spain at 10.1 million tonnes, which means that the output, subject to a different pattern of vegetables, is over one quarter higher per hectare in Greece than in Spain. This is brought about by the greater emphasis in Greece of vegetables grown for their fruit, 2.9 million tonnes.

Fruit Production in Greece

Fruit production by volume in Greece is 2.7 million tonnes or 32 per cent of that in Spain. The fruit cropping of Greece is given in Table 12. The schedule of crops is matched for type of crop without exception to the fruit cropping schedule for Spain. Of apples and pears, which comprise 11 per cent of the fruit area in Greece, pears are over a quarter, in line with their importance in other Mediterranean countries. Stone fruit is prominent in Greece and at 21 per cent of the fruit area, is the same proportion as that under stone fruit in Spain. Of the 49 thousand ha of stone fruit in Greece over two thirds, no less than 30 thousand ha, are in peaches, so different from the fruit cropping in Italy and Spain. Apricots and cherries in Greece are each some six thousand ha. Soft fruit, also not very prominent in a Mediterranean climate are only between one and two per cent of the fruit area in Greece. Of the three thousand ha of soft fruit in Greece, kiwifruit dominate with over two thousand ha. Eurostat classes kiwi fruit as soft fruit. Citrus fruit is only 24 per cent of the area of all fruit in Greece. This compares with citrus at 42 per cent of the fruit area in Spain. Among the citrus fruit, oranges are more important in Greece, at two-thirds, whereas in Spain oranges are a little over half the area of citrus fruit. Among the other fruit in Greece ten thousand ha, or three quarters of the other fruit class, is down to figs. Figs are also prominent in the fruit cropping in Spain. In Greece, fresh grapes are more prominent than in Spain. Greece has 81 thousand ha as compared with Spain's 68 thousand ha.

Vegetable Exports in Greece

Greek exports of vegetables are given in Figure 7 and Table 13. In Greece an overwhelming proportion of the vegetable exports, 89 per cent, go to West Germany. Nearly ten per cent go to other

Table 11. Greece - Vegetable Production, Areas and Quantities,1987

Items as listed in Eurostat

	'000 ha	'000 tonnes
Cauliflowers	3.1	54.3
Cabbage	8.5	183.0
All brassicas	**11.6**	**237.3**
Celeriac and celery	0.6	13.3
Leeks	1.7	42.5
Lettuce	3.0	64.5
Endive	2.0	36.8
Spinach	2.7	43.0
Asparagus	2.1	7.0
Globe artichoke	2.9	28.2
Other vegetables (leafed or stalked)	2.0	42.0
Vegetables leafed or stalked	**17.0**	**277.3**
Tomatoes	38.1	1665.1
Cucumbers	2.9	147.9
Gherkins	0.8	12.9
Melons	8.9	149.6
Water melons	18.1	627.0
Aubergines	3.0	74.7
Pumpkins, marrows	4.7	95.5
Peppers	4.2	98.4
Other cultivated for fruit	2.2	18.0
Cultivated for fruit	**82.9**	**2889.1**
Carrots	1.0	32.1
Garlic	2.2	14.5
Onions	11.7	179.4
Beetroot	0.7	17.2
Radish	0.4	6.0
Roots and tubers	**16.0**	**249.2**
Peas	2.8	16.6
French beans and runner beans	7.7	72.0
Other pulses	1.4	10.9
Pulses	**11.9**	**99.5**
Total of vegetables specified	**139.4**	**3752.4**

Source: Eurostat 1989

Table 12. Greece - Fruit Production, Areas and Quantities, 1987

Items as listed in Eurostat

	'000 ha	'000 tonnes
Fresh grapes	**81.0**	**722.0**
Apples	17.5	303.2
Pears	6.9	119.9
Apples and pears	**24.4**	**423.1**
Peaches	33.0	547.5
Apricots	6.4	106.7
Cherries	6.2	34.5
Plums	0.6	2.7
Nectarines	2.3	27.9
Other stone fruit	0.1	3.1
Stone fruit	**48.6**	**722.4**
Strawberries	0.7	5.7
Kiwi	2.2	2.5
Other soft fruit	0.1	2.3
Soft fruit	**3.0**	**10.5**
Oranges	34.8	578.9
Mandarins	4.6	46.6
Satsumas	.0	3.5
Clementines	.0	20.7
Lemons	12.6	118.5
Grapefruit	0.3	4.8
Other citrus fruit	0.3	2.9
Citrus fruit	**52.6**	**775.9**
Figs	9.6	42.3
Quinces	0.3	7.7
Avocados	0.2	0.3
Other fruit NOS	2.7	15.9
Other fruit	**12.8**	**66.2**
Total of fruit specified	**222.4**	**2720.1**

Source: Eurostat 1989

Figure 7. Greek Exports of Vegetables by Country of Destination

Exports by Quantity - 1987

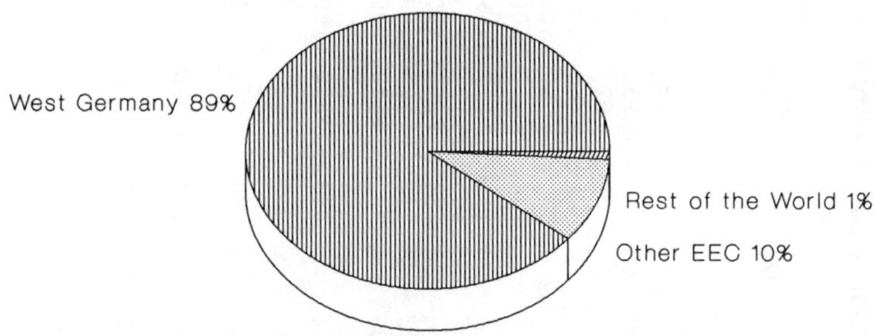

Source: Eurostat External Trade, Exports, 1989

Exports by Value - 1987

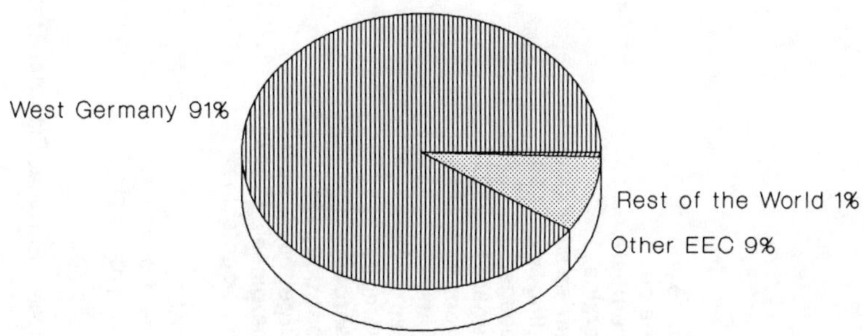

Source: Eurostat External Trade, Exports, 1989

Table 13. Greek Exports of Vegetables by Country of Destination - 1987 tonnes

	TOTAL	UK	WG	FR	BL	NE	DE	IT	EEC	PER CENT	INTRA EEC	OTHER COUN-TRIES	PER CENT
Lettuce	5											5	100.0
Tomatoes	216	1	137	20					158	73.1	159	58	26.9
Cucumbers	44316	467	39201	376	3424	731	53		44252	99.9	44252	64	0.1
Gherkins	6977		6977						6977	100.0	6997		
Other salad vegetables	1	1							1	100.0	1		
Cauliflowers	90		90						90	100.0	90		
Cabbages	10		10						10	100.0	10		
Celeriac	18		18						18	100.0	18		
Carrots	4											4	100.0
Onions	363		24	86			230	23	363	100.0	363		
Leeks	141					125			125	88.7	126	16	11.3
Asparagus	4530		3975			535			4510	99.6	4510	20	0.4
Mushrooms	1											1	100.0
Sweet peppers	12330		11718			115		293	12126	98.3	12127	204	1.7
Courgettes	1	1							1	100.0	1		
Aubergines	133	2	2	16					20	15.0	20	113	85.0
Other vegetables	953	31	718	6	1	85	2	76	919	96.4	918	34	3.6
TOTAL	70089	503	62870	504	3425	1591	285	392	69570		69592	519	
Percent specified country of total		0.7	89.7	0.7	4.9	2.3	0.4	0.6	99.3		99.3	0.7	

Source: Eurostat External Trade, Exports, 1989

Figure 8. Greek Exports of Fruit (excluding citrus) by Country of Destination

Exports by Quantity - 1987

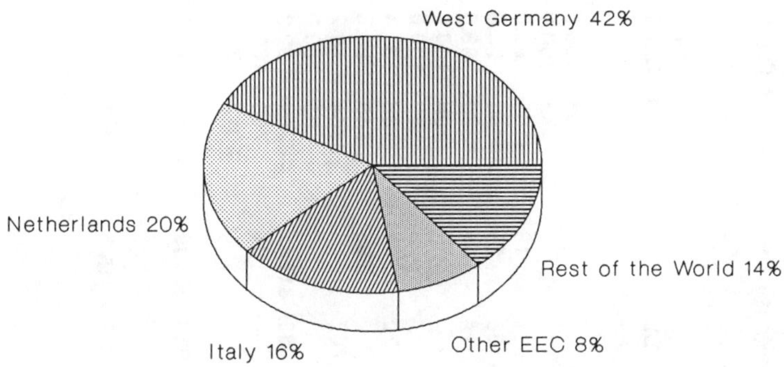

Source: Eurostat External Trade, Exports, 1989

Exports by Value - 1987

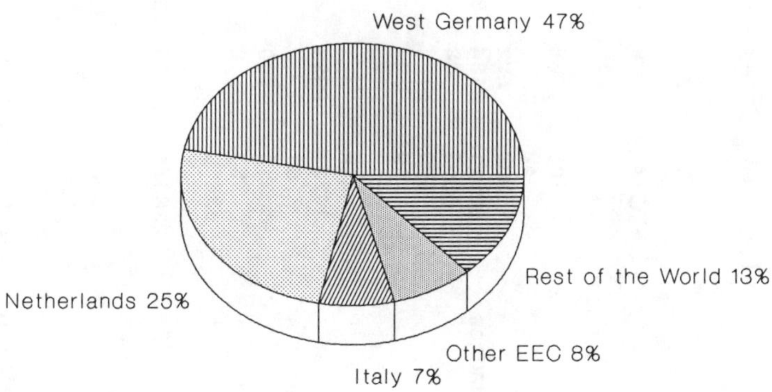

Source: Eurostat External Trade, Exports, 1989

Table 14. Greek Exports of Fruit (excluding citrus) by Country of Destination, 1987, tonnes

	TOTAL	UK	WG	FR	BL	NE	DE	IT	EEC	PER CENT	INTRA EEC	OTHER COUN-TRIES	PER CENT
Bananas	106	58	32			16			106	100.0	106		
Avocados	42		16	9		17			42	100.0	41		
Apricots	15909	592	7094	589	159	5703		18	14155	89.0	14206	1754	11.0
Peaches and nectarines	60111	2134	32533	160	32	7954	1188	39	44040	73.3	44040	16071	26.7
Cherries	8362	48	3459			4749	14	59	8329	99.6	8349	33	0.4
Plums	109		15			94			109	100.0	109		
Other stone fruit	89		44			45			89	100.0	89		
Apples	8727	18	760		12	520	70	5723	7103	81.4	7103	1624	18.6
Pears	446		150					295	445	99.8	473	1	0.2
Table grapes	90781	8833	46341	289	91	24633	767		80954	89.2	81107	9827	10.8
Other fresh berries	53		27	26					53	100.0	53		
Water melon	50217	2100	8705	708	1069	1626	900	31241	46349	92.3	46349	3868	7.7
Other fresh melon	763	2	130	21		576		34	763	100.0	763		
Other fresh fruit	2856		1077			935			2012	70.4	2012	844	29.6
TOTAL	238571	13785	100383	1802	1363	46868	2939	37409	204549	85.7	204800	34022	14.3
Percent specified countries of total		5.8	42.1	0.8	0.6	19.6	1.2	15.7	85.7			14.3	

Source: Eurostat External Trade, Exports, 1989

countries in the EEC and only one per cent to countries in the rest of the world. West Germany takes a total of over 0.6 million tonnes, of which cucumbers alone are 0.4 million. Apart from cucumbers, the biggest exports to West Germany are sweet peppers, 12 thousand tonnes, gherkins, 7 thousand tonnes and asparagus, 4 thousand tonnes. Tomato exports from Greece are very low at 216 thousand tonnes. Belgium-Luxembourg, take 4.9 per cent of the Greek export of vegetables, which is virtually all cucumbers. Holland takes 2.3 per cent, principally cucumbers 731 tonnes, asparagus 535 tonnes and leeks, 125 tonnes.

A little over 500 tonnes of vegetables are exported to "other countries". This one per cent of vegetable exports goes to four countries only. Sweden takes the most, and seven vegetables are included in the exports to Sweden. Austria is next with three items of export, tomatoes, cucumbers and asparagus. Finland and Switzerland each take one tonne of mushrooms and certain other vegetables. It is clear that European countries, with little fruit and vegetable production of their own, rely on countries other than Greece for their supply of vegetables. Greece does not export any vegetables to countries in Eastern Europe which lie close to Greece. The total absence of vegetable exports from Greece to countries in Eastern Europe is rather a surprise, and it means that the people in these countries only consume limited amounts of exotic vegetables.

Greek Exports of Fruit other than Citrus Fruit

Greek exports of non-citrus fruit are given in Table 14 and illustrated in Figure 8. West Germany also takes the largest share of Greek exports of non-citrus fruit, at 42 per cent, not an overwhelming share as in the Greek exports of vegetables. In the EEC, the Netherlands comes next with 20 per cent, followed rather surprisingly by Italy with 16 per cent. Other countries in the EEC take only eight per cent . These are led by the UK with six per cent. Unlike vegetable exports where the exports to the rest of the world is only one per cent, the rest of the world takes 14 per cent of the non-citrus Greek fruit. The composition of Greek exports of non-citrus fruit is determined by a few important fruits. For instance, table grapes are 38 per cent of the exports, peaches and nectarines are 25 per cent and water melons are 21 per cent. Apricots are 6.7 per cent of the non-citrus fruit exports. Of the table grapes, well over half go to West Germany, but none to the other Mediterranean exporters, Spain, Italy and Portugal. West Germany also takes well over half of the Greek export of peaches and nectarines, of which only 18 tonnes go to Italy. The 14 per cent to Italy is of some interest. Of the 37 thousand tonnes of non-citrus fruit going to Italy, 31 thousand are water melons and almost six thousand tonnes are apples. It is difficult to find reasons for this trade other than traditional business ties between Greece and Italy, and the high quality standard of the fruit concerned. Twenty seven per cent of the peaches and nectarines go to other countries than the EEC. The most, four thousand tonnes go to Austria but three thousand tonnes go to Czechoslovakia and over two thousand tonnes go to Belgium and Luxembourg. It would seem that some luxury fruit is being taken up by countries in Eastern Europe but the market would seem to be limited to a select few. Saudi Arabia takes three thousand tonnes and Kuwait, two thousand. Of the eleven per cent of the Greek exports of table grapes, the largest single export of some ten thousand tonnes, goes to countries outside the EEC. Over nine thousand tonnes go to Switzerland and smaller quantities to Sweden and Finland. Most of the nearly four thousand tonnes of water melons go to Austria and the balance to Sweden. In total Austria takes eight per cent of the Greek non-citrus fruit exports, more than the countries in the EEC, with the exception of West Germany and Italy. The small tonnage of bananas come from Crete.

Greek Exports of Citrus Fruit

Greece exports 0.24 million tonnes of citrus fruit. There are several features of interest and the most striking one is the high proportion of citrus fruit exported to countries in Eastern Europe and Soviet Russia. Another point is that Greek citrus exports are almost entirely oranges and lemons. Twenty seven per cent of the Greek exports of citrus goes to all the EEC countries together, of which 12 per cent goes to West Germany, as shown in Figure 9 and Table 15. A little over eight per cent go to the Netherlands, nearly four per cent goes to Britain and some three per cent goes to other countries in the EEC. In addition, of the almost 0.2 million tonnes of oranges exported, somewhat less than

Figure 9. Greek Exports of Citrus Fruit by Country of Destination

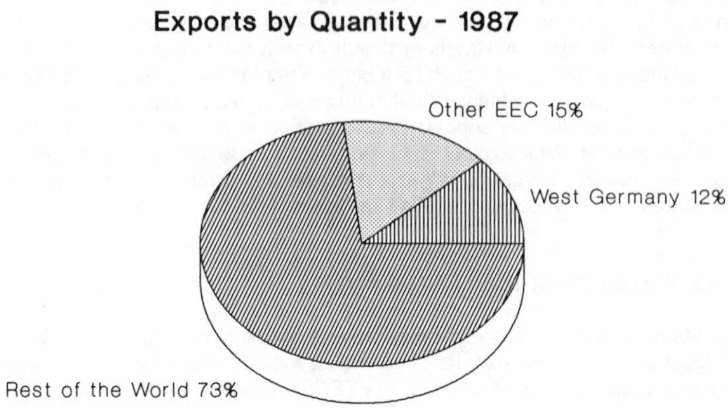

Source: Eurostat External Trade, Exports, 1989

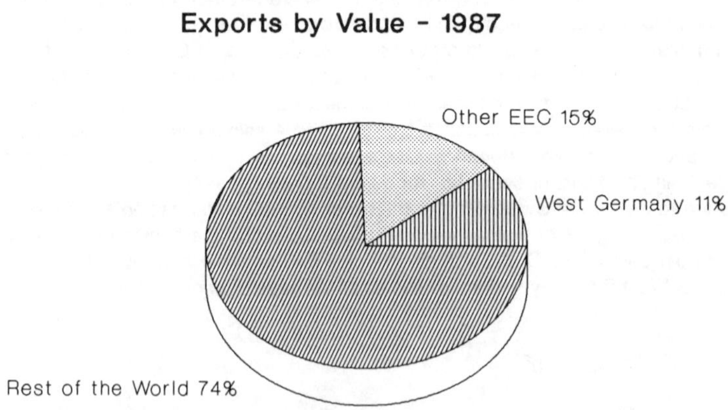

Source: Eurostat External Trade, Exports, 1989

Table 15. Greek Exports of Citrus Fruit by Country of Destination - 1987 tonnes

	TOTAL	UK	WG	FR	BL	NE	DE	EEC	PER CENT	INTRA EEC	OTHER COUN-TRIES	PER CENT
Oranges	194918	8669	28015	2662	1604	19559	2938	63447	32.6	265	131471	67.4
Other sweet oranges	70			70				70	100.0	70		
Sanguines	146					146		146	100.0	146		
Clementines	75	16	18	5			13	52	69.3	52	23	30.7
Mandarins and Wilkins	37		34	3				37	100.0	37		
Other citrus hybrids	12			12				12	100.0	12		
Lemons	45360	358	315				1	674	1.5	675	44686	98.5
Grapefruit	166	151		10			5	166	100.0	166		
TOTAL	240784	9194	28382	2762	1604	19705	2957	64604	26.8	1423	176180	73.2
Percent specified countries of total		3.8	11.8	1.1	0.7	8.2	1.2	26.8			73.2	

Source: Eurostat External Trade, Exports, 1989

a third go to EEC countries and somewhat more than two-thirds go to "other countries", mainly countries in Eastern Europe and the Soviet Union. Of the 45 thousand tonnes of lemons exported, 98.5 per cent goes to other countries than the EEC, also mainly to East Europe and Soviet Russia. The biggest export market is the Soviet Union which takes ten per cent of the oranges and 36 per cent of the lemons, a total export market share of 14.5 per cent. Poland has a 14 per cent market share, a little less than Soviet Russia. These exports are almost entirely lemons. The next largest market in Eastern Europe is Rumania, which takes 11.3 per cent. The Rumanian market comprises virtually all oranges, amounting to 27 thousand tonnes and lemons only 18 thousand tonnes. East Germany, with a 9.5 per cent market share, like Rumania, takes over 20 thousand tonnes of oranges and one thousand tonnes of lemons. Czechoslovakia has a 7.5 per cent market share, the majority of which are oranges. Yugoslavia has a 5.4 per cent market share, comprising nearly 11 thousand tonnes of oranges and two thousand tonnes of lemons. Bulgaria has a 3.4 per cent market share, all of which is oranges. Hungary also has a 3.4 per cent market share, also all oranges.

Outside Eastern Europe, Austria has a 4.3 per cent market share of Greek exports of citrus fruit, again almost all oranges. Very small quantities of citrus fruit go to Norway, Sweden and Finland who get most of their supplies from Italy.

The discrete market for Greek citrus fruit shows that Eastern Europe and Soviet Russia are getting some citrus fruit that they cannot provide for themselves. In the kitchens of Eastern Europe, lemons, like garlic, are a necessity. There is a high level of fruit production in Hungary, Rumania and Bulgaria with a favourable climate for fruit other than citrus. However, while some peaches are grown, citrus fruit is not grown. Apart from small imports from Turkey, it is unlikely that East Europe and Soviet Russia are importing from any other source, and it is believed that the southern Soviet states will have little citrus fruit to export in the Soviet internal market. This means that Greek exports of citrus fruit to East European countries and Soviet Russia is a very big and important part of their total supply. That Greek citrus fruit exports go to East Europe and the Soviet Union, while vegetable and other fruit exports do not. suggests that East Europe and Soviet Russia are managing with the vegetables and other fruit which they themselves grow. There are also important and regular exports of tropical fruit to Soviet Russia and Eastern Europe from Cuba. There is no doubt a strong tradition, over generations, of trade in citrus fruit between Greece, and the Eastern European countries and Soviet Russia, and geographical proximity is important. There can be no doubt, that Greek exports of citrus fruit to Eastern Europe and Soviet Russia is more profitable to Greece than alternative markets. This situation pertains, of course, to the great quantity of vegetables, some ninety per cent of Greek vegetable exports, which go to West Germany. The considerable exports of Greek vegetables to West Germany illustrates a profitable market for Greek vegetables.

The exports of citrus fruit from Greece to Eastern Europe and the Soviet Union in total are 166 thousand tonnes. This compares with imports of citrus to West Germany alone of one million tonnes, which means that Eastern Europe and Soviet Russia are taking very little imports of citrus fruit for the population they represent. The market for citrus fruit imports to Eastern Europe will increase, as incomes rise, to meet consumer preferences. Hopefully the situation in Soviet Russia will improve. These markets for citrus fruit so very small at present may be expected to grow immensely in future.

PORTUGAL

The Portuguese Production of Vegetables

Portuguese data on fruit and vegetables is not as well developed as that in other Member States. Portuguese vegetable production is shown in Table 16. There is a total listed of 87 thousand hectares of vegetables with a production of 1.76 million tonnes. Brassicae crops are a little over ten per cent of the vegetable area. Vegetables grown for their fruit are substantially tomatoes, which are 20 thousand ha in the total of 87 thousand ha. The area of lettuce grown in Portugal is one thousand ha and that of roots and tubers is 5 thousand ha. Total production in tonnes of vegetables in Portugal is seventeen per cent that of Spain, which means that the Portuguese stake in international trade in vegetables is lower.

Portuguese Exports of Vegetables

Portuguese exports of vegetables amount to only 5.7 thousand tonnes. These exports are given in detail in Table 17 and they are illustrated in Figure 10. The EEC takes 86 per cent, and important trading partners within the EEC are the Netherlands, 29 per cent, the UK, 25 per cent, and France 21 per cent. Other EEC countries take 11 per cent of Portugal's vegetable exports between them. The two largest exports of vegetables are lettuce and onions, each some 1300 tonnes. Tomato exports are just over one million tonnes, in the total of vegetable exports of 5.7 million tonnes. Mushrooms are relatively important among Portuguese exports of vegetables and prices for mushrooms are good in comparison with the price of other vegetables.

Of exports of vegetables to countries outside the EEC, 93 tonnes of lettuce, a small fraction of Portuguese exports, and 15 tonnes of cucumbers go to Sweden. The Ivory Coast takes 400 tonnes or seven per cent of Portuguese onion exports. A token amount of "other vegetables" and of tomatoes go to the USA.

The Portuguese Production of Fruit

As given in Table 18, the total area of fruit in Portugal covers only 90 thousand hectares and the volume of produce is 414 thousand tonnes. One third of the fruit area is apples and pears, of which a high proportion is pears. Apples and pears is almost one third of the total fruit production. The quantity of stone fruit is prominent with an area of 19 thousand ha, and a production of 64 thousand tonnes. Peaches comprise half the Portuguese stone fruit area, while cherries are next in prominence at nearly four thousand ha. Apricots and plums are each 2.8 thousand ha and 2.5 thousand ha. The total production of stone fruit, 63.9 thousand tonnes, is 15 per cent of the total fruit production. Soft fruit has little significance in Portugal. Of the 2.8 thousand ha of soft fruit, as much as 1.7 thousand ha is not specified.

Citrus fruit occupies a large proportion of the fruit area in Portugal, at thirty five per cent of the total area of fruit production of 90 thousand ha. The proportion of citrus fruit, by value, is the same as the proportion by area, at 36 per cent of the total. Oranges occupy 62 per cent of the area under citrus and they account for 71 per cent of the production of citrus fruit by volume. Mandarins, at nearly seven thousand ha, occupy 21 per cent of the area under citrus fruit, and they account for 16 per cent of the citrus produced. Lemons are a third of citrus fruit, by area, at 4.8 thousand ha or 15 per cent of the total citrus area. Production by volume of lemons is 18 thousand tonnes, 42 per cent of the total fruit production. Grapefruit, as in other Mediterranean exporting countries, is very low in production.

Table 16. Portugal - Vegetable Production, Areas and Quantities, 1987

Items as listed in Eurostat

	'000 ha	'000 tonnes
Cauliflowers	1	19
Other brassicas	8	147
All brassicas	**9**	**166**
Lettuce	1	30
Vegetables leafy or stalked	**1**	**30**
Tomatoes	20	670
Melons	3	20
Water melons	0	2
Peppers	0	1
Cultivated for fruit	**23**	**693**
Carrots	3	81
Onions	2	65
Roots and tubers	**5**	**146**
French beans and runner beans	3	28
Pulses	**3**	**28**
Other fresh vegetables	**45**	**697**
Total of vegetables specified	**87**	**1,760**

Source: Eurostat 1989

Figure 10. Portuguese Exports of Vegetables by Country of Destination

Exports by Quantity - 1987

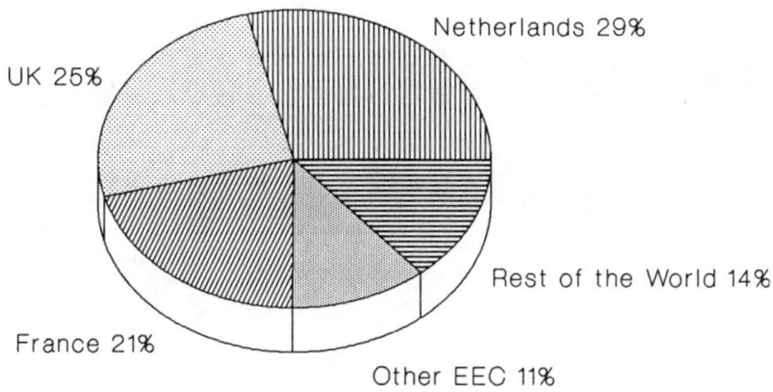

Source: Eurostat External Trade, Exports, 1989

Exports by Value - 1987

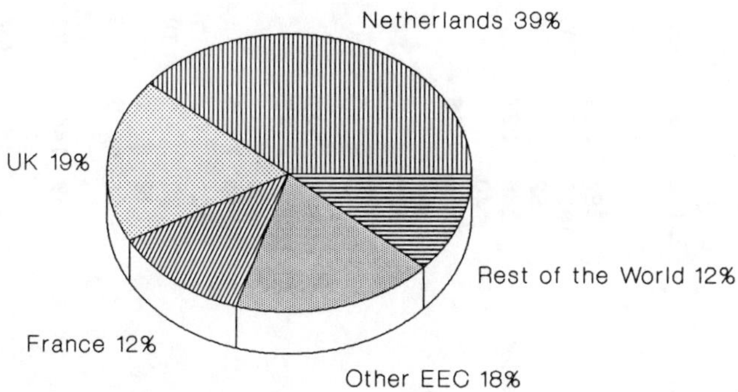

Source: Eurostat External Trade, Exports, 1989

Table 17. Portuguese Exports of Vegetables by Country of Destination - 1987 tonnes

	TOTAL	UK	WG	FR	BL	NE	DE	SP	IT	EEC	PER CENT	INTRA EEC	OTHER COUN-TRIES	PER CENT
Lettuce	1397	761				442	29			1232	88.2	1233	165	11.8
Tomatoes	1068	38	291	77	23	632				1061	99.3	1060	7	0.7
Cucumbers	62		3				44			47	75.8	47	15	24.2
Other salad vegetables	246	227				19				246	100.0	246		
Cauliflowers	40			8		32				40	100.0	40		
Cabbages	212	16	1	164	1	30				212	100.0	207		
Peas	68			51		17				68	100.0	70		
Beans	503		9	77		417				503	100.0	504		
Other legumes	30					30				30	100.0	30		
Carrots	1			1						1	100.0	1		
Edible roots	327	327								327	100.0	327		
Onions	1246	43		664	114	15		10		846	67.9	846	400	32.1
Garlic	6				6					6	100.0	6		
Leeks	10	10								10	100.0	10		
Mushrooms	401		1	123					76	200	49.9	19	201	50.1
Sweet peppers	29	1	10	16		2				29	100.0	403		
Aubergines	2	2								2	100.0	2		
Other vegetables	41	16	8			2			11	37	90.2	38	4	9.8
TOTAL	5689	1441	323	1181	144	1638	73		87	4897	86.1	5089	792	13.9
Percent specified country of total		25.3	5.7	20.8	2.5	28.8	1.3	0.0	1.5	86.1			13.9	

Source: Eurostat External Trade, Exports, 1989

Table 18. Portugal - Fruit Production, Areas and Quantities, 1987

Items as listed in Eurostat

	'000 ha	'000 tonnes
Fresh grapes	**6.1**	**60.3**
Apples	19.2	87.5
Pears	11.0	45.5
Apples and pears	**30.2**	**133.0**
Peaches	9.7	28.0
Apricots	2.8	6.8
Cherries	3.8	8.6
Plums	2.5	8.5
Nectarines		12.0
Stone fruit	**18.8**	**63.9**
Strawberries	0.8	2.4
Kiwi	0.3	0.1
Other soft fruit	1.7	6.1
Soft fruit	**2.8**	**8.6**
Oranges	20.5	105.0
Mandarins	6.7	23.8
Lemons	4.8	18.3
Grapefruit	0.2	0.9
Citrus fruit	**32.2**	**148.0**
Total of fruit specified	**90.1**	**413.8**

Source: Eurostat 1989

Figure 11. Portuguese Exports of Fruit (excluding citrus) by Country of Destination

Exports by Quantity - 1987

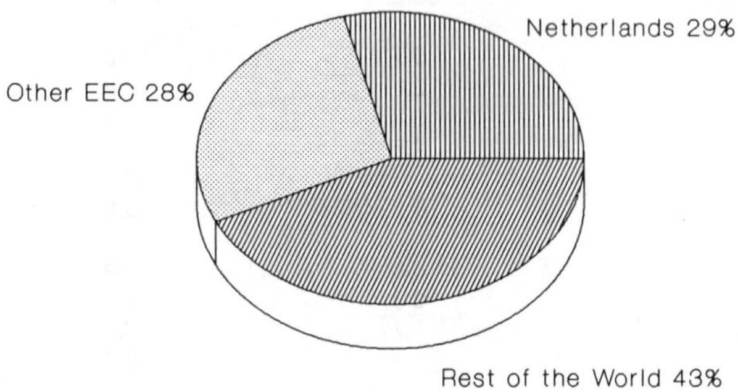

Source: Eurostat External Trade, Exports, 1989

Exports by Value - 1987

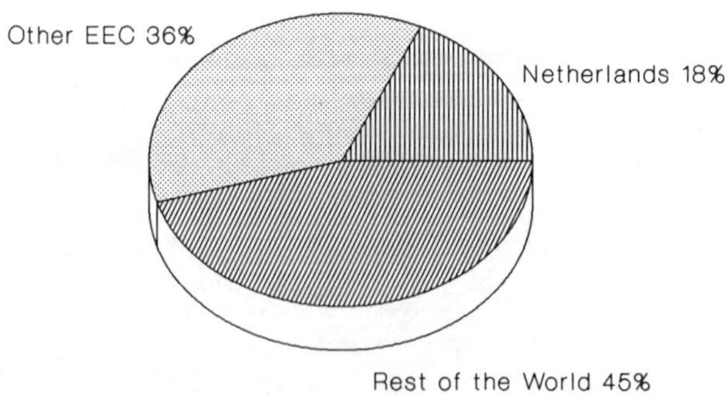

Source: Eurostat External Trade, Exports, 1989

Table 19. Portuguese Exports of Fruit (excluding citrus) by Country of Destination 1987, tonnes

	TOTAL	UK	WG	FR	BL	NE	DE	IT	IR	SP	EEC	PER CENT	INTRA EEC	OTHER COUN-TRIES	PER CENT
Kiwifruit	354				6					348	354	100.0	59		
Apricots	23			23							23	100.0	46		
Peaches and nectarines	155			112	42	1					155	100.0	167		
Cherries	11				11						11	100.0	11		
Apples	3223	18				2319	6	18	90	18	2469	76.6	6905	754	23.4
Pears	3330	40		37			10			251	338	10.2	337	2992	89.8
Plums	15	1		5	9						15	100.0	15		
Table grapes	156	37	75	27	17						156	100.0	156		
Strawberries	868	492	300	11	1	5	2				811	93.4	500	57	6.6
Other fresh berries	61			61							61	100.0	61		
Water melon	19			17	2						19	100.0	19		
Other fresh melon	648	214	154	2	39	239					648	100.0	647		
TOTAL	8863	802	529	295	127	2564	18	18	90	617	5060	57.1	8923	3803	42.9
Percent specified countries of total		9.0	6.0	3.3	1.4	28.9	0.2	0.2	1.0	7.0	57.1			42.9	

Source: Eurostat External Trade, Exports, 1989

Exports of Fruit other than Citrus Fruit

Figure 11 illustrates the Portuguese export trade in non-citrus fruit, and Table 19 gives this in more detail. Portuguese trade in non- citrus fruit indicates that Portugal has a more international trade. Trade in fruit, excluding citrus fruit, in Portugal has a different pattern from that of Spain, Italy and Greece. Twenty-one per cent of Portuguese exports in these commodities goes to the Netherlands, an almost identical amount goes to other countries in the EEC, and 43 per cent goes to the rest of the world. This trade pattern is different from the other Mediterranean exporters, and it is most likely due to traditional trade links going back for a century or more.

Only two commodities, apples and pears, each with some three thousand tonnes of exports, dominate the exports from Portugal of non-citrus fruits. Exports of stone fruit of all kinds, which for peaches and nectarines are the highest at 155 tonnes and go to France and Belgium, are in all other cases much less than the small export of peaches and nectarines. The consignment of kiwifruit, 354 tonnes, substantially goes to Spain. This is a commodity which Spain could easily produce for herself. After the export of strawberries of 868 tonnes, of which over half go to the UK and nearly thirty per cent to West Germany, melon exports are the most important at some 650 tonnes. These go mainly to the UK, the Netherlands and West Germany. By far the largest consignment of apples, 72 per cent, go to the Netherlands. The 3330 tonnes of pears are exported to Brazil, 1639 tonnes and Canada, 1353 tonnes, with smaller amounts to Spain, Denmark and the UK. Some 750 tonnes of apples go to Brazil, while Sweden, Switzerland and Austria take small quantities of strawberries.

Portuguese Exports of Citrus Fruit

The pattern of the Portuguese export trade in citrus fruit, shown in Figure 12 and Table 20, is unique to Portugal. France takes no less than 92 per cent of Portuguese exports of citrus fruit. This is explained by the tradition of sea trade at a distance up the French coast, from exporting ports in Portugal, both linked by a common sea. The almost one hundred per cent Portuguese sales of citrus fruit to France, recalls that West Germany takes virtually ninety per cent of the vegetable exports of Greece. The very small exports of citrus fruit to the UK and Belgium, and the very high French trade, make up the total of Portuguese citrus fruit exports.

Table 20. Portuguese Exports of Citrus Fruit by Country of Destination, 1987, tonnes

	TOTAL	UK	FR	BL	EEC	PER CENT	INTRA EEC
Oranges	12		12		12	100.0	12
Other sweet oranges	14		14		14	100.0	14
Sanguines	181		181		181	100.0	379
Clementines	456		456		456	100.0	456
Other citrus hybrids	12			12	12	100.0	12
Lemons	33	2	31		33	100.0	33
Grapefruit	2		2		2	100.0	2
TOTAL	710	2	696	12	710	100.0	
Percent specified countries of total		0.3	98.0	1.7	100.0		

Source: Eurostat External Trade, Exports, 1989

Figure 12. Portuguese Exports of Citrus Fruit by Country of Destination

Exports by Quantity - 1987

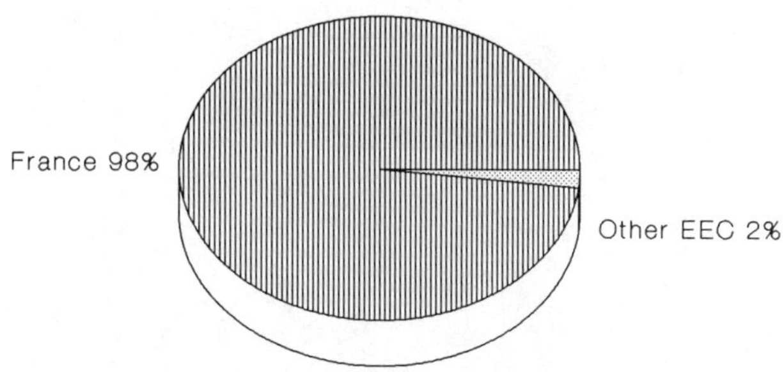

Source: Eurostat External Trade, Exports, 1989

Exports by Value - 1987

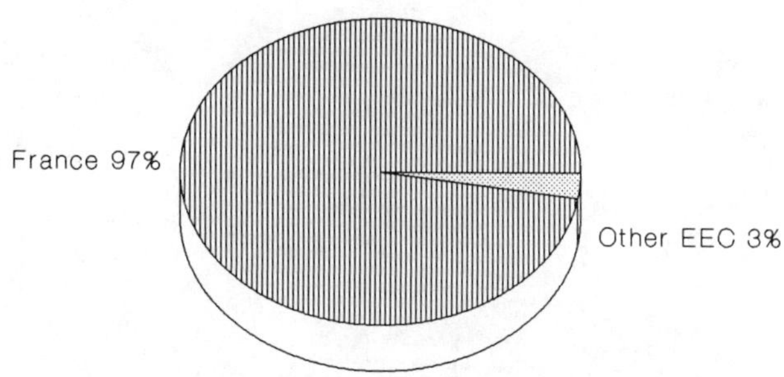

Source: Eurostat External Trade, Exports, 1989

The composition of the exports of citrus fruit for Portugal is interesting. Oranges, so prominent in production, are a minor export, as are other than sweet oranges. The largest export is of clementines, 456 tonnes, followed by sanguines, 181 tonnes. There are few exports of lemons and yet fewer exports of grapefruit. Obviously it is policy in Portugal to export a substantial amount of clementines and sanguines, but, as with vegetable produce in Portugal, the domestic market is important. Trade along the Atlantic seaboard, and tradition in this trade, accounts for the distinctive pattern of the Portuguese trade in fruit and vegetables. Trade flows of the four Mediterranean exporting countries do suggest commercial linking of a most competitive nature, but at the same time, trade flows in practice are affected by the different and important trade routes. Savings in transport costs are seen to be supported by the logic of different routes for the different countries and their markets, which have been developed over time. With this, the commercial benefit to both sides in the fruit and vegetable trade has also grown.

A Comparison of the Mediterranean Countries

A comparison of the export values of the Mediterranean countries is given in Table 21 for vegetables, non-citrus fruit and citrus fruit. The impact of Spain and Italy as important exporters is shown. Italy is relatively low for citrus fruit. Greece is very much an exporter of non-citrus fruit, and Portugal is hardly in the market at all and the contribution of Portugal to the EEC market is low.

Table 21. Comparison of Export Values of the Mediterranean Countries, 1987 - million ECUs

		Intra EEC	Other Countries	World
Spain	Vegetables	465.6	55.5	537.0
	Non-citrus fruit	378.0	70.5	456.1
	Citrus fruit	816.3	149.0	971.1
Italy	Vegetables	328.5	80.0	419.6
	Non-citrus fruit	818.9	158.6	989.6
	Citrus fruit	30.7	41.5	75.7
Greece	Vegetables	43.7	0.2	44.3
	Non-citrus fruit	104.7	16.0	123.0
	Citrus fruit	14.2	41.2	57.1
Portugal	Vegetables	5.0	0.7	6.3
	Non-citrus fruit	2.6	2.2	5.1
	Citrus fruit	0.3	0.0	0.3

Source: Eurostat External Trade, Exports,1989

CHAPTER 8

THE OPPORTUNITY FOR DEVELOPING COUNTRIES IN THE EUROPEAN MARKET FOR FRUIT AND VEGETABLES

This chapter includes a feature on the international structure of Developing Country trade in fruit and vegetables. Secondly it deals with the contribution of agricultural development relating to the fruit and vegetable enterprise. There follows a series of market studies which all indicate that there is scope for developing countries to send more produce to Europe at a profit. Then there is a section on EEC trade policy as a handicap to the developing countries' trade, and one on the prospects for developing countries trade. Finally there is a discussion of the banana sector and reference to the situation for fruit and vegetables in the Single Market of 1992.

The International Structure of Developing Country Trade in Fruit and Vegetables

This section relies heavily on data from the FAO Trade Yearbook[1] where South Africa, Australia and New Zealand are treated as developed countries. To put developing country trade in perspective with trade in agriculture and related commodities, world exports of these commodities are given in Table 1. Certain commodities in the processed form are included because of their importance to developing countries.

Exports of Agriculture, Fish and Forestry Produce - Table 1

This shows exports of agriculture, fish and forestry products (column 1), agricultural products (column 2) and food, excluding fish (column 3). Column 1 shows the share of developing country exports in million dollars to be 28 per cent of world exports and the share of developed country trade to be 72 per cent. As would be expected, the share of agricultural, fish and forestry products exports from developing countries is low, at little over a quarter, but of course the share of total exports of developing countries is substantially less. Agricultural commodities are important in the exports of developing countries. Between regions, of the developing countries total, the exports of Latin America and the Far East dominate, being 40 and 39 per cent respectively.

Agricultural Exports

The pattern of agricultural exports, excluding fish and forestry products (column 2) is broadly similar to those in column 1, but the proportion of exports of agricultural products from developing

---

[1] FAO Trade Yearbook, Vol. 41. Rome, 1988.

Table 1. Developing Country Exports of Agricultural Products, 1987 and Percentage of World Exports

	Column 1 Agriculture, Fish and Forestry Products			Column 2 Agricultural Products			Column 3 Food, excluding Fish		
	million dollars	per cent world exports	per cent regions	million dollars	per cent world exports	per cent regions	million dollars	per cent world exports	per cent regions
World exports	3429867	100.0		2512357	100.0		1622065	100.0	
Developing market economies	835762	24.4		646694	25.7		363208	2.4	
Africa	106765		12.8	88070		13.7	48272		13.3
Latin America	337306		40.3	291827		45.1	174712		48.1
Near East	57253		6.8	54973		8.5	37909		10.4
Far East	324132		38.8	206368		31.8	98787		27.2
Other	10305		1.3	5456		0.9	3528		1.0
Centrally planned economies	263318	7.7		186300	8.0		113082	7.0	
Asian	119491		45.4	89531	48.1		55410		49.0
Eastern Europe and USSR	143827		54.6	96770	51.9		55672		51.0
All developed economies	2474615	72.1		1776132	70.7		1203460	74.2	
All developing economies	955253	27.9		736224	29.3		418619	25.8	

Source: FAO Trade Yearbook, Vol. 41, 1988.

countries is a little higher at 29 per cent and those of developed countries are 71 per cent. While exports of agricultural products from the developing regions are still dominated by Latin America and the Far East, Latin America is yet more prominent with exports at 45 per cent, and the share of the Far East is lower at 32 per cent. The share of Africa is up from 13 per cent in column 1 to 14 per cent, while that of the Near East is up from seven to eight and a half per cent.

With food, excluding fish, isolated in column 3, the developing world exports are less prominent than in the case of exports of agricultural, fish and forestry products and of agricultural products alone. The share of developing country exports drops to 26 per cent and that of developed country exports rises to 74 per cent. In food exports Latin America is even more prominent with 48 per cent and the share of the Far East drops markedly to 27 per cent. The share of Africa is only a little less than for agricultural products, but the share of the Near East is higher at just over ten per cent.

Potatoes, pulses and tomatoes - Table 2

Coming on to fruit and vegetables and the developing country share in world trade, Table 2 provides data for potatoes, pulses and tomatoes. The DITC number is given in brackets in each case. These three commodities show marked contrasts in the share of developing country trade. The share of potatoes of developing countries in world exports is 15 per cent, of pulses a high 38 per cent, and of tomatoes, it is 22.5 per cent. Distance, and the weight of a comparatively low value commodity, account for the low percentage of potatoes in world exports, while the keeping quality of pulses and their relatively high value account for their prominence. Associated with this is the fact that pulse crops are important in tropical regions. Potato exports from the developing regions are dominated by the near East (65.8 per cent), which exports new potatoes to Europe and elsewhere. Next in prominence, though a long way behind, is Africa, with 16.5 per cent. Potato exports from Latin America and the Far East are each about half those of Africa. The exports of pulse crops are also dominated by the Near East. Latin America pulse exports are a quarter of the regional exports of developing countries but the Far East has 21.5 per cent and Africa just over ten per cent. Pulse exports are thus of importance to all the developing regions of the world. The regional pattern of tomato exports is very different. As much as 63.5 per cent of the tomatoes come from Latin America, with the market in the United States close at hand, and having advantages in climate in the off-season of supply to the United States. Exports from the Far East are only marginal, while those to Africa and the Near East are 17 per cent and 19 per cent respectively.

World exports of potatoes are worth 1088.5 million dollars, that of pulses is 1833.5 million dollars and of tomatoes 1472 million dollars. The world share of centrally planned economies in this trade is low, as it is in agricultural products right across the board. It is 3.7 per cent for potatoes, 10.3 per cent for pulses and 5.8 per cent for tomatoes. Eastern Europe and the USSR are most prominent in the centrally planned economies' exports of potatoes and tomatoes, whereas Asian sources are important for pulses, in line with the prominence which the Near East has in developing country exports of pulses.

Apples, Pears and Peaches - Table 3

Developing country exports of apples, pears and peaches are given in Table 3. World exports of apples are worth 1557.5 million dollars. Pears and peaches are worth much less, 452 million dollars and 537 million dollars respectively. This reflects the durability and the demand for apples and to a lesser extent for pears. Trade in apples and pears is substantially inter-hemispherical to catch the off-season of supply in the affluent northern regions of the world. Peaches are by far the most perishable of these fruits, which is reflected in the low percentage export trade, 8.3 per cent of world exports of developing countries, so many of which are distant from their markets, which often implies a long and time consuming journey. The share of developing country exports in apples and pears is much higher at 23 and 25 per cent. Apples and pears are ideally a crop of the temperate climate, where they benefit from winter dormancy. Thus exports from Africa are negligible, while those of Latin America dominate at 76 per cent of developing country exports. In Latin America there is scope for gaining production advantages in altitude which gives a cooler climate. The export trade in apples of the Near and Far East is 16 and eight per cent respectively of all developing country exports. Centrally planned economies' export trade is low at four per cent and this is mainly accounted for by Eastern Europe

Table 2. Developing Country Exports 1987 and Percentage of World Exports

	Potatoes (054.1)			Pulses (054.2)			Tomatoes (054.3)		
	thousand dollars	per cent world exports	per cent regions	thousand dollars	per cent world exports	per cent regions	thousand dollars	per cent world exports	per cent regions
World exports	1088558	100.0		1833496	100.0		1471918	100.0	
Developing market economies	157452	14.5		58645	31.9		329891	22.4	
Africa	25859		16.5	60187		10.3	55621		16.9
Latin America	15378		9.8	149201		25.4	209130		63.5
Near East	103393		65.8	249764		42.8	62727		19.0
Far East	12321		7.9	126348		21.5	2043		0.6
Other							10		
Centrally planned economies	40203	3.7		189230	10.3		85738	5.8	
Asian	8044		20.0	110050		58.2	2109		2.5
Eastern Europe and USSR	32159		80.0	79180	41.8		835629		97.5
All developed economies	923063	84.8		1137804	62.1		1139918	77.5	
All developing economies	165496	15.2		695695	37.9		332000	22.5	

Source: FAO Trade Yearbook, Vol. 41, 1988.

Table 3. Developing Country Exports 1987 and Percentage of World Exports

	Apples (057.4)			Pears (057.92)			Peaches (057.93)		
	thousand dollars	per cent world exports	per cent regions	thousand dollars	per cent world exports	per cent regions	thousand dollars	per cent world exports	per cent regions
World exports	1577567	100.0		451616	100.0		537070	100.0	
Developing market economies	336622	21.0		87335	19.3		40257	75.0	
Africa	8								
Latin America	255825		76.0	69242		79.3	300		0.7
Near East	53058		15.8	2656		3.0	33501		83.3
Far East	27731		8.2	15437		17.7	6433		16.0
Other							23		
Centrally planned economies	132116	3.9		27493	6.1		10716	20.0	
Asian	28780		21.8	25019		91.0	4200		39.2
Eastern Europe and USSR	103336		78.2	2474		9.0	6516		60.8
All developed economies	1212165	76.8		339262	75.1		492613	91.7	
All developing economies	365402	23.2		112354	24.9		44457	8.3	

Source: FAO Trade Yearbook, Vol. 41, 1988.

and the USSR.

Since the climatic requirements for pears and the capacity for storage are similar to that for apples the pattern for pear exports is similar, with Latin America even more dominant with 79 per cent. The proportion of pear exports of the Far East at 18 per cent, however, compares with that for apples of eight per cent, and the pear export share of the Near East falls to three per cent as opposed to that of apples of 16 per cent. The dominant developing country region for peaches with 83 per cent, is the Near East, which has a warmer climate than is ideal for apples and pears, but which is most suitable for peaches. Also the Near East is closer to markets in Europe. Africa does export a small quantity of peaches. The centrally planned economies are much more prominent in the export share of peaches, 20 per cent of the world export trade; 61 per cent are traded by Eastern Europe and the USSR and 39 per cent are traded by Asian centrally planned economies.

Grapes, Raisins and Dates - Table 4

Grapes are important in world exports of fruit and vegetables. Table 4, which gives the exports of fresh grapes, also includes the dried grape or raisin, and dates. World exports of grapes are 1146 million dollars, those of raisins 653 million dollars, and those of dates 150 million dollars. The export share of developing countries in the world exports of grapes is relatively high at 30 per cent. Latin America dominates the trade with 85 per cent, followed a long way behind by the Near East at 13 per cent. No exports are recorded for Africa where it is generally too hot, and a small percentage of developing country exports is shown for the Far East. Raisins, or dried grapes, are not perishable like fresh grapes and they tell a different story. The developing country share of world exports of raisins is 43.5 per cent. Ninety-two per cent come from the Near East where the climate is ideal for grapes, 6.5 per cent from Latin America and a very small amount from the Far East.

Dates, because of their climatic requirements and because of their durability in transport, are very much traded by developing countries, such that 77.5 per cent of the world exports of dates come from developing economies. Africa is prominent in the trade in dates with 38 per cent of the developing economy trade. The Near East is more important with 46 per cent, the balance being made up by the Far East with 16.5 per cent. The centrally planned economies have five per cent of the world exports of grapes, which are nearly all from Eastern Europe and the USSR. Centrally planned economies' exports of raisins are negligible, and the Asian economies account for only 0.9 per cent of world trade in dates.

Oranges, Lemons and Limes - Table 5

Table 5 shows developing country exports of three groups of citrus fruit: oranges, lemons and limes, and other citrus fruit. The share of the exports of developing countries is 27 per cent for oranges and 21.5 per cent for each of the other two commodities. For oranges, Africa exports the largest percentage, 36 per cent, but the Near East and Latin America closely follow with 31 and 29 per cent respectively. Lemon and lime exports are largest in the Near East with 60 per cent. They are 38 per cent for Latin America and one per cent only in Africa and the Far East. Other citrus fruit is featured strongly in Latin American trade with 75 per cent of developing country regional exports. The Near East exports are 17 per cent, with Africa six per cent, and the Far East, ten per cent.

World exports of oranges are large, 2360 million dollars, appreciably more than the other large export, bananas, while those of lemons and limes and other citrus are much smaller, between four and five hundred million dollars. Like date exports, Africa is prominent in the orange trade.

Bananas, Coconuts and Desiccated Coconuts - Table 6

As with dates, bananas, coconuts and desiccated coconut, featured in Table 6, are also very much the realm of developing country production for export. Developing countries have 92 per cent of world banana exports, 84 per cent of coconut exports and 95 per cent of desiccated coconut exports. Bananas, with oranges, are the foremost fruit in consumption in the markets of the developed world and world exports are worth nearly 2,000 million dollars. Coconuts are much less important with exports of 20 million dollars but world exports of desiccated coconut are 141 million dollars. Banana exports are dominated by Latin America at 89 per cent, with seven per cent from the

Table 4. Developing Country Exports 1987 and Percentage of World Exports

	Grapes (057.51)			Raisins (057.52)			Dates (057.96)		
	thousand dollars	per cent world exports	per cent regions	thousand dollars	per cent world exports	per cent regions	thousand dollars	per cent world exports	per cent regions
World exports	1145940	100.0		652900	100.0		149846	100.0	
Developing market economies	345368	30.1		283403	43.4		115173	76.9	
Africa							43551		37.8
Latin America	292784		84.8	18502		6.5			
Near East	45854		13.3	261476		92.3	52667		45.7
Far East	6750		1.9	3243		1.2	18955		16.5
Other									
Centrally planned economies	56851	5.0		800	0.1		1025	0.9	
Asian	1920		3.4						
Eastern Europe and USSR	54931		96.6	800		100.0			
All developed economies	798652	69.7		36897	56.5		33648	22.5	
All developing economies	347208	30.3		284200	43.5		116198	77.5	

Source: FAO Trade Yearbook, Vol. 41, 1988.

Table 5. Developing Country Exports 1987 and Percentage of World Exports

	Oranges (057.1)			Lemons and Limes (057.21)			Other Citrus (057.22/29)		
	thousand dollars	per cent world exports	per cent regions	thousand dollars	per cent world exports	per cent regions	thousand dollars	per cent world exports	per cent regions
World exports	2359893	100.0		419295	100.0		452885	100.0	
Developing market economies	593079	25.2		90286	21.5		139915	30.9	
Africa	215269		36.3	970		1.1	7965		5.6
Latin America	172424		29.1	34326		38.1	105276		75.3
Near East	183321		30.9	54029		59.8	23535		16.8
Far East	22055		3.7	948		1.1	3137		2.3
Other									
Centrally planned economies	48542	2.1		58			2826	0.6	
Asian	47442		97.7	58			2826		100.0
Eastern Europe and USSR	1000		2.3						
All developed economies	1715372	72.8		328951	78.5		310144	68.5	
All developing economies	640521	27.2		90344	21.5		90344	21.5	

Source: FAO Trade Yearbook, Vol. 41, 1988.

Table 6. Developing Country Exports 1987 and Percentage of World Exports

	Bananas (057.3)			Coconuts (057.71)			Coconuts (desiccated) (057.71)		
	thousand dollars	per cent world exports	per cent regions	thousand dollars	per cent world exports	per cent regions	thousand dollars	per cent world exports	per cent regions
World exports	1950663	100.0		19765	100.0		141538	100.0	
Developing market economies	1734723	88.9		16693	84.4		135067		
Africa	54673		3.2	1450		8.7	2700		2.0
Latin America	1545072		89.1	6249		37.4	1377		1.0
Near East	6955		0.4				168		0.1
Far East	126703		7.3	8659		51.9	130422		96.9
Other	1320			335		2.0			
Centrally planned economies	58681	3.0							
Asian	58681								
Eastern Europe and USSR									
All developed economies	157259	8.0		3072	15.6		6471	4.6	
All developing economies	1793404	92.0		16693	84.4		135067	95.4	

Source: FAO Trade Yearbook, Vol. 41, 1988.

Far East and three per cent from Africa. The Asian countries of the centrally planned economies account for a small percentage of world banana exports.

The export from developing countries of coconuts and desiccated coconut has a very different pattern from bananas. Of developing country exports of coconuts the Far East has a little over half and Latin America more than a third, followed by Africa with nine per cent, but practically all, 97 per cent, of desiccated coconut is exported from the Far East.

Fresh Pineapples, Canned Pineapples and Pimento - Table 7

The world export trade in fresh pineapples reaches a little over two hundred million dollars in value, and the trade in canned pineapples is more than twice this amount. Again, whilst this text does not deal with processed fruit, it is useful to give the trade in tinned pineapples. Pineapples are very much a developing country export with over 80 per cent of each type coming from developing countries. Africa dominates the supply of fresh pineapples from developing countries with 55 per cent, Latin America has 25 per cent and the Far East 20 per cent. The exports of canned pineapples, like dried coconuts, is however dominated by the Far East, with 84 per cent. Africa has 12 per cent and Latin America four per cent. Figures for pimento, given in Table 7, shows that pimento is a majority developing country export with 61 per cent, and 63 per cent of the developing regions exports come from the Far East. Next in prominence is Latin America with 27 per cent and ten per cent comes from Africa. It reaches the market fresh or dried, with most of the production dried.

Vanilla, Peppers and Onions - Table 8

Vanilla and pepper are also prominent in the export trade of developing countries with 61 per cent of world exports of vanilla and 91 per cent of world exports of pepper. These high percentages illustrate the durability of these products in transport and storage as well as their relatively high value. Again, because of the importance of vanilla and pepper to developing countries, these two commodities are included in this analysis. A large proportion of world exports of vanilla comes from the centrally planned economies, 49 per cent, with the three-quarters from the Asian regions and a quarter from Eastern Europe and the USSR. Among developing market economies with 61 per cent of world exports, 86 per cent comes from Africa and 24 per cent from the Far East. Centrally planned economies also export 19 per cent of world exports of pepper. Among the developing market economies the Far East exports over three quarters of pepper exports and Latin America exports 22 per cent.

Onions are produced over a great range of climates in the world. The United Kingdom regularly gets imports of onions from a number of countries and the edge on quality and freshness of onions produced in a warmer climate over those from Britain and Holland, justifies the expense of bringing in onions from great distances. These onions fetch a premium price. Developing countries account for a relatively high world export figure Of 33 per cent. All principal regions of the developing market economies export onions. The Far East is again the most prominent with 40 per cent, but the Near East send 35 per cent and Latin America, 21 per cent. Centrally planned economies feature with eight per cent of world exports of onions, most of which come from Eastern Europe and the USSR.

All developing regions benefit from world trade in fruit and vegetables, and it is quite clear from the foregoing analysis of the structure of developing country exports of agricultural and fruit and vegetable products, that developing countries have an important role in the world exports of these products, but the level of this activity for each commodity depends on the nature of the individual commodity concerned. It is high for commodities which need a hot climate, but it is also a fact that most developing countries in the world are in the hot climate belts. It is their advantage in climate which gives such commodities their strength in international trade in fruit and vegetables. Again, in general, the most successful commodities in trade are those which can stand the long journey from the country of origin to the markets, and which are of sufficient value to offset the cost of transport. Processed commodities, favoured by climate in production, also have the advantage of providing more employment in the producing developing country concerned, than growing and marketing only.

Table 7. Developing Country Exports 1987 and Percentage of World Exports

	Pineapples (Fresh) (057.95)			Pineapples (Canned) (057.95)			Pimento (075.1)		
	thousand dollars	per cent world exports	per cent regions	thousand dollars	per cent world exports	per cent regions	thousand dollars	per cent world exports	per cent regions
World exports	206221	100.0		423281	100.0		133881	100.0	
Developing market economies	162139	78.0		346098	81.7		38934	29.1	
Africa	88650		54.8	41465		11.9	4025		10.3
Latin America	40114		24.7	14217		4.1	10345		26.6
Near East	6						78		0.2
Far East	33319		20.5	290416		84.0	24421		62.7
Other	50						65		0.2
Centrally planned economies	8027	3.9		14175	4.1		58279	48.5	
Asian				14175		100.0	42700		73.3
Eastern Europe and USSR	8027		100.0				15579		26.7
All developed economies	36055	17.5		63008	14.9		52240	39.0	
All developing economies	170166	82.5		360273	85.1		81634	61.0	

Source: FAO Trade Yearbook, Vol. 41, 1988.

Table 8. Developing Country Exports 1987 and Percentage of World Exports

	Vanilla (075.21)			Pepper (075.1)			Onions (054.51)		
	thousand dollars	per cent world exports	per cent regions	thousand dollars	per cent world exports	per cent regions	thousand dollars	per cent world exports	per cent regions
World exports	120078	100.0		664147	100.0		467267	100.0	
Developing market economies	103726	86.4		591570	89.1		147986	31.7	
Africa	89470		86.3	7441		1.3	4351		2.9
Latin America	760		0.1	129110		21.8	31658		21.4
Near East	78			1924		0.3	52500		35.5
Far East	24421		23.5	453095		76.6	59461		40.2
Other	65						16		
Centrally planned economies	58279	48.5		12691	19.1		37237	8.0	
Asian	42700		73.3	12558		99.0	3979		10.7
Eastern Europe and USSR	15579		26.7	113		1.0	33258		89.3
All developed economies	52247	39.0		60019	9.0		315302	67.4	
All developing economies	81634	61.0		604128	91.0		151965	32.6	

Source: FAO Trade Yearbook, Vol. 41, 1988.

A Summary of the Major Fruit and Vegetables in International Trade

Developing countries have a major share of world export trade in several commodities which are normally grown in the tropics. These commodities are dates, bananas, coconuts and desiccated coconuts, pineapples and canned pineapples, pimento, vanilla and pepper. Developing countries have a 44 per cent share in the world export trade in raisins, but the other commodities in which up to 30 per cent of the exports come from developing countries are also of great importance to the developing economies of developing countries. These are potatoes, 15 per cent, pulses, 38 per cent, tomatoes, 23 per cent, apples, 23 per cent, pears, 25 per cent, peaches, 8.3 per cent, grapes, 30 per cent, oranges, 27 per cent, lemons and limes, 22 per cent and other citrus fruits, 22 per cent.

Bananas are second only to oranges in world consumption. Nearly 90 per cent of developing country exports of bananas come from Latin America. By contrast 85 per cent of fresh coconuts are exported by developing countries and 52 per cent of these come from the Far East. Developing countries account for 95 per cent of world exports of dried coconuts, and 97 per cent of these come from the Far East. So all regions of the developing world share in world trade in fruit and vegetables. This is much to their advantage in gaining export earnings and in developing their economies. Some countries depend on export earnings from fruit and vegetables because they have little else to export.

It is a handicap to developing countries that the markets in the developed economies are so very much protected, and it is important for levels of protection to be reduced. It is a fortunate coincidence that many agricultural products exported from developing countries are not produced in the countries of the developed world. Protection, and therefore interference with free trade, is most advanced for the domestically produced produce of the advanced economies.

The Conditions for Agricultural Development

Schermerhirn gives a considered analysis of problems and prescriptions in the production and marketing of fruit and vegetables in developing countries. To start with, one must look at the systems of fruit and vegetable production in developing countries, in a sector to play its part in economic development.[2] Of course the fruit and vegetables must be marketable at a price to give a margin over the costs of production, and this is especially so in the long run. An infra-structure must be established or be in place to move the produce from the production area to the market. The infrastructure in developing countries will be the one which needs attention. In some third world countries, marketing of fruit and vegetables has been considered as a way of disposing of a surplus. Such an approach is, of course, not appropriate. What is needed is a market-orientated approach to production. This merely means that production capabilities, resources and facilities must be co-ordinated and planned to produce what is needed by the consumers at a price which can be afforded by them. This requires that agricultural producers must respond to consumer demand rather than merely attempting to sell what they produce. This seems to be obvious but producers in developing countries are often handicapped by a poor level of education such that it is important to adopt a strict discipline to production and to production techniques and quality standards. This applies in particular to the grading operation. At the same time the developing country producer has a strong advantage over the developed country producer. Wages are considerably less and no social costs have to be met by developing country employers. As a proprietor, the new developing country producer of fruit and vegetables would also have a lower expectation of income generated by the business.

The Role of Government

The role of government is usually one of the most basic and important ways of establishing fruit and vegetable enterprises. The step from a role of a peasant type of person, to become a successful fruit or vegetable producer, is a very big one. The new fruit or vegetable producer needs training as a producer, and given technical advice as well as advice on marketing. The comprehensive training of

---

[2] A Comparative Analysis of Fruit and Vegetable Marketing in Developing Countries, by R.W. Schermerhirn, The Post-harvest Institute for Perishables, University of Idaho, Sigma One Corporation, 1986.

the producer and his being launched into very competitive exporting business, requires a strong input by the appropriate government department, or of a special marketing board or like institution.

The Contribution of the Marketing System

A marketing system can constitute a contribution to the development in the following ways. It can minimise transaction and exchange costs between producers and consumers, while it can add to producer incomes. Market information, which makes producers more aware of supply and demand conditions and product movements and prices, can reduce risks in the business. Another contribution of marketing systems is that it can also provide methods which bring subsistence farmers into an exchange economy by providing arrangements for collective action, and introduce technology appropriate to marketing on a small scale. Further, the marketing system can reduce post-harvest losses. The marketing system also provides more employment in the economy. These needs are basic in developing countries, and, as has been said, the fruit and vegetable sector may be the first to provide employment. The conditions necessary for effective marketing include support services and various institutions and activities and programmes conducted by the private or the public sector. These are designed to create a climate conducive to an efficient marketing system for fruit and vegetables. The necessary support services are research, extension, market intelligence and information, market regulation and financing. Effective marketing using extension programmes must provide information in the following areas. These will include the best alternatives in marketing, pricing, outlook and technical information which will assist participants in the marketing system. Finally, what is needed is the availability and use of harvest and post-harvest technology, including grades and standards to ensure consumers receive the quality products they want and to minimise post-harvest losses. The quality standards of the European Economic Community must be achieved in a fairly primitive environment. This is no mean task for workers or proprietors in a developing country where the usual attitude to quality is much less exacting. So far as finance is concerned, credit schemes aimed at increasing the production of fruit and vegetables are common in agricultural development programmes. Finance aimed at marketing is less common and in many cases does not exist. As a fruit and vegetable marketing system develops, financing may be needed during the performance of any of the marketing functions, such as grading, packaging and transport, storage. Financing may be needed by an individual or it may be needed to assist a group of farmers.

The Potential Market for Exotic Fruit and Vegetables from Developing Countries

There have been numerous papers on the market for exotic produce from tropical developing countries over the last five years. Perhaps the most important one is the paper by Proctor given at the recent ISHS Symposium on Tropical Fruit in international trade.[3] The EEC market for tropical fruit has grown rapidly from 1983 to 1987. For the major fruit, namely avocados, mango and pawpaws, market growth has doubled to give a total of 390 thousand tonnes in 1987, from 1983. Proctor gives a comprehensive classification of the EEC market in tropical fruits for the major produce listed in Table 9. The expensive commodities, mangoes, avocados and pineapples are widely known and they have a firm place in the market and the greatest expansion in demand.

Indeed in Japan, perhaps the most important market for tropical fruit, the practise has developed of the family being offered slices of mango, for instance. The Japanese have raised exotic fruit eating as a cult, as important as the drinking of alcohol.

Pineapple consumption in the EEC has doubled in volume over the five years. Imports of pine apples to certain countries in the EEC for 1987, in thousand tonnes, is as follows. Imports to France are 66 thousand tonnes, to Italy 45,000, to West Germany 39 thousand, and to Belgium- Luxembourg 28 thousand tonnes. Over 70 per cent of the exports of pineapples to the EEC come from the Ivory Coast. Other sources include Costa Rica, Ghana, the Philippines, Cameron and South Africa. The major exporters offer a year round supply to Europe.

---

[3] The European Community Market for tropical fruit and the factors limiting their growth. Acta Horticurae Proctor F.J. No 269 pp 29-40 1990.

Table 9. A Classification of tropical fruit and import volume (tonnes) and value (ECU) into the EEC in 1983 and 1987

Group	Fruit	1983 tonnes	1983 ECU '000	1987 tonnes	1987 ECU '000
Well established (15 years or more)	Pineapple	99880	82414	238936	158009
	Avocado	71149	101394	115654	128444
Moderately well established (5-15 years)	Mango[1]	12895	24049	28011	44868
	Pawpaw	1358	2945	5074	9142
Recently established	"Other fruit"[2]	49668	100181	35710	46781
	"Other berries"[3]	9757	5720	13143	10620

(1) Includes guavas and mangosteens.
(2) "Other fruit" includes cherimoyas, passion fruit and physalis. From 1987 "Other fruit" also includes soursop, custard apple, granadilla, loquats, sapodillas, lychees, rambutans, pulasans.
(3) "Other berries" includes breadfruit, babacos, carambolas, durian, feijoas, longans, jackfruit, lychees, persimmons, pomegranates, prickly pears, sapodillas and tamarillos.

Source: Acta Horticulturae No 268 The European Market for tropical fruit and the factors limiting its growth.

While avocado is one of the most successful of the sub-tropical fruits marketed in recent years, the EEC represents less than ten per cent of the world market. Of the 115 thousand tonnes going to the EEC, France is now the major market in Europe, taking 65 per cent. The UK market is second, some 19 thousand tonnes. West Germany comes next with only ten thousand tonnes, and the Netherlands takes four thousand. Avocados are available throughout the year. The Israel season is from October to May with its peak in November to April. Avocados arrive from South Africa, March to December with a peak May to September, the United States from November to July, and then Spain and the Canary Islands from October to April. There is a long season of supply, and by marketing a subtropical variety such as Hass, Fuerte and Ettinger which are adapted to sea transport and managed marketing, the growth and potential for avocado is considered to be large. When there is a significant production of avocados in the EEC some third countries can be expected to lose their market.

The mango is the most important tropical fruit which is moderately well established in the European market. In 1987 the UK took 9.6 thousand tonnes, France, six thousand and West Germany and the Netherlands four thousand tonnes. From 1983 to 1987 a three-fold growth in the market took place. The market growth for mango has been limited by the large number of suppliers and the relatively short season of availability from any single source, and by the lack of technology for consistent and reliable transport by sea. With current sources of supply, there is a wide range of varieties on the market. The strongest demand is for the Florida types, however specific market outlets, mainly associated with the ethnic minority groups, insist on non-Florida types, for example the Asian and the Caribbean varieties. Exports to the EEC of pawpaws, in 1987, despite an increase in consumption in the EEC, was only five thousand tonnes. The major markets are West Germany, 1362 tonnes, and the UK. Brazil supplies over 30 per cent. Other sources include Costa Rica and the Ivory Coast. It is said that over twenty tropical fruits are offered on the market on a regular basis. The more common of this group include lychee, passion fruit, caramola, guava, physalis, mangosteen and cherimoya. These fruits are the subject of increasing interest and curiosity and for some the market will increase over the next few years.

It is pointed out that the retail outlets need a continuity of produce throughout the year. A short season of two to three months for mangoes does not make the business viable to the retailer. Establishing horticultural sector involves many problems and risks. The levels of production, post-harvest technology and marketing applied to home and regional industries are extremely different to those applied to international trade, and as such means that an export programme requires specific support inputs. This may include financial investment, which can be high, and such expenditure must be critically assessed to evaluate the merits of the development or expansion of an enterprise.

Tropical and Off-season Fresh Fruit and Vegetables

The publication by International Trade Centre on the European import markets for off-season fresh fruits and vegetables gives a good illustration of the market for the main fruit and vegetables, which developing countries have the climatic potential to produce for the European market. In Table 10 commodities in the top half of the table come mainly from developed countries at present, while most commodities in the lower half of the table, except peaches and nectarines, come from developing countries. The fruit in largest supply is peaches and nectarines at almost half a million tonnes, fol lowed by capsicums at 0.3 million tonnes. Imports from 0.1 to 0.3 million tonnes in order of their volume are melons, pineapples, water melons and avocados. The smallest import listed is of pawpaws at less than three thousand tonnes.[4]

West Germany is by far the most important market with the highest market share for apricots, asparagus, capsicums, water melons, pawpaws and peaches and nectarines. France has the second place in the market share with the largest market share for aubergines, courgettes, avocados, mangoes and pineapples, and the second largest market share is for apricots, capsicums, water melons and pawpaws in order of volume. The market share of the UK is less, but Britain has the highest market share for mangoes and melons. The markets of West Germany, France and the UK, represent the largest market for off-season fresh fruits and vegetables. The import market for ten European countries is listed in Table 10. From 1982 to 1986 the ten European markets have expanded by seventy-five per cent for avocados, 98 per cent for pineapples and 130 per cent for mangoes. Only a few years ago these fruits were still considered luxury items, imported in small quantities and sold at high prices into the better restaurants and speciality shops. For pineapples and avocados the expansion is explained by the availability of refrigerated sea transport and the reduction of transport costs, and to Europeans travelling overseas and to promotion of this produce. Mangoes and pawpaws are still air freighted so they are comparatively expensive. Promotional campaigns carried out by supermarket chains and by the EEC, through Comitè de liaison des fruit tropical et legumes de contre - saison ordinaires des Etats ACP (COLEACP) have decidedly influenced consumption and the development of certain markets. It is also true that in some countries certain ethnic groups, to whom mango and pawpaws are common fruits, already account for a significant share in total demand and consumption.

The main obstacle to a further increase in the sales of tropical fruits in general is the lack of knowledge on the part of the consumer. However, there is a growing interest in most European countries, and if prices can be reduced by improved organisation of production and marketing, and by the development of sea freight, the effects of well planned promotion campaigns should lend to a continual climb in consumption. The consumption of avocados in Western Europe started increasing in the 1970's and it is still growing. Imports to Western Europe were 105 thousand tonnes in 1986. France is by far the largest market for avocados followed by Britain and West Germany. The leading exporters of avocados are Israel, South Africa, Spain and Martinique. Kenya, with Mexico and the United States come some way behind. European imports of avocados are higher from October to April than in the summer months when supplies of fruit grown in Europe are available.

[4]Tropical and Off-Season Fresh Fruit and Vegetables, International Trade Centre, Geneva 1987

Table 10. Imports of Off-Season Fresh Fruits and Vegetables into selected European Markets, 1986 (total tonnage and percentage each country)

	Apricots	Asparagus	Aubergines	Avocados	Capsicums	Courgettes
Tonnes	75388	31539	47315	104715	302986	81762
	%	%	%	%	%	%
W. Germany	37.5	69.6	28.4	7.5	46.6	12.5
France	20.8	1.7	41.4	63.7	17.5	60.0
UK	4.1	4.1	11.0	14.2	10.9	15.8
Netherlands	4.3	0.8	5.8	3.4	8.4	2.5
Sweden	0.1	0.2	n.a	3.1	3.9	n.a
Switzerland	15.1	17.7	3.7	2.8	3.8	6.2
Austria	12.1	1.2	4.9	1.0	3.6	n.a
Denmark	0.1	0.2	1.2	1.9	2.0	0.5
Norway	n.a	n.a	n.a	0.5	n.a	n.a
Belgium Luxembourg	5.9	4.5	3.6	1.9	3.3	2.5

	Mangoes*	Melons	Water melons	Pawpaws	Peaches & nectarines	Pineapples
Tonnes	23931	214643	168193	2857	496283	170679
	%	%	%	%	%	%
W. Germany	12.9	13.1	47.5	30.7	47.7	21.4
France	25.8	4.6	31.6	16.1	9.1	39.6
UK	36.9	37.7	9.0	23.8	16.0	11.2
Netherlands	18.6	12.9	4.9	16.7	5.4	11.3
Sweden	2.0	6.2	n.a	n.a	2.5	1.0
Switzerland	n.a	9.5	n.a	n.a	2.5	1.0
Austria	n.a	7.7	n.a	n.a	4.3	1.7
Denmark	1.1	1.2	2.1	1.9	1.9	1.5
Norway	n.a	n.a	n.a	n.a	n.a	1.4
Belgium Luxembourg	52.7	7.1	4.9	10.8	6.6	3.9

*Mangoes, Mangosteens and Guavas

Source: Tropical and Off-Season Fresh Fruits and Vegetables, International Trade Centre, Geneva 1987

Chilies are a more pungent variety of capsicum used by the immigrant population, especially people of Mediterranean or Asian origin. Ginger arrives from the United States, from Kenya and the Caribbean from July to December and from Brazil from December to May. The containerisation of limes by sea is growing. Lychees and rambutans arrive in Europe from late November to early March. Britain has the largest market share of mangoes and France has the second largest market share.

Mangoes are usually air freighted. For okra there is a steadily growing European market which suggests opportunities for further consumption in the winter months. France is the biggest importer of pineapples followed by West Germany and the UK. Imports are concentrated between October and May, with a peak in December. Western Europe is the second market after Japan, and Spain and Italy have a considerable demand for pineapples, which imported 16,000 and 36,000 tonnes respectively in 1985. Pineapples come mainly from the Ivory Coast, but the demand for other exotic fruit in Spain and Italy is very small. The European off-season is the full European winter, from October to April, when little fresh produce can be grown in the open. In practise the actual off-season depends on the product concerned, in particular the period when southern hemisphere and tropical produce is available. This stems from the fact that Europeans have expected off-season produce to be available over the year, and that they can afford to buy it.

The actual timing of the European demand for individual produce will of course vary with the produce and the country or origin. The European off-season is said to be becoming reduced by the effect of southern European producers using plastic cover. However, it is more than likely that the market for off-season tropical produce in Western Europe will increase at a faster rate. After all, the South European producers can only grow a small proportion of the tropical fruits and vegetables since the South European climate is not warm enough to produce fruits and vegetables of a truly tropical origin. At the same time the data on imports of tropical fruit to Western Europe shows that the total demand for this produce is expanding now, and it will expand more in future. The precise size of this demand will depend on income levels in Europe, and on the success of promotion campaigns to make the exotic attributes of these fruit and vegetables better known and understood by the consumers.

The report by Joy covers the market for 67 fresh commodities in four EEC member states the UK, France Western Germany and the Netherlands. In the main the commodities were originally imported for ethnic minority groups, people of African, Asian and Caribbean descent. The indigenous populations now are increasingly inclined to include this exotic produce in their diet. For most of the commodities covered, developing countries are the main source of supply but certain countries and regions tend to dominate in a commodity or group of commodities, or become a source of supply to a particular EEC country. The supply to individual EEC countries has largely evolved from historical ties and it is further maintained by the availability of produce. At present the fruit and vegetables covered mentioned are still minor items in the fruit and vegetable trade from developing countries, but in future this group of lesser known fruit and vegetables is expected to achieve growth rates above the average level expected for established produce, so they will increase in relative importance. The fruit and vegetables with the best immediate prospects in the four countries studied are mango, pawpaws, persimmon, (the less astringent variety), lychee and to a lesser extent passion fruit and okra. Other produce, highlighted as having prospects in one or two of the four EEC countries, are lime, kumquat, guava, cherimoya and sweet potato. All these items have already made an impact in the market.[5]

The UK is the only important market for all starchy root crops but imports are declining. France is a major market for sweet potato. A number of items are consumed almost wholly by the minor ethnic groups. Some limited increase is now to be expected in this sector. Prospects will depend on whether or not the indigenous population start to consume these products.

Compared with volume lines such as citrus fruit and European grown fruit and vegetables, the fruit and vegetables were expensive and this may limit demand. While the minority ethnic groups are prepared to pay high prices to eat their traditional foods, consumers in the indigenous market are disinclined to pay high prices. Lower prices are needed if the demand is to grow significantly, but this will be difficult to achieve in the near future. Most of the produce has to be air-freighted, which is expensive and costs from 30 to 60 per cent of the wholesale price, and the cost is rising. Prospects are enhanced for those items which potentially can be sent by sea and this situation is being actively investigated. For each ethnic group, the exception being West Germany, there are separate trading channels for exotic fruit, which are distinct from the customary trading channels for the indigenous population. These trading channels serving the ethnic groups remain strong because so many of these items are still destined for these groups. Growing interest by the indigenous population in some

---

[5]Selected European Markets for Speciality and Tropical Fruit and Vegetables, 1987, Joy, Corinne

of these lesser known items has been taken up by importers servicing this group, who now import a selected range of produce. Immediate prospects for suppliers lie with the indigenous market sector and for the restricted range of items. Even for these items, and with the exception of mangoes, market development will be achieved only slowly. Also new suppliers will come to compete with existing suppliers depending on the period of supply.

The United Kingdom Market for Fresh Exotic Fruit

Diversification of exports into fruit and vegetables can yield a number of economic benefits for developing countries. These include increased and more stable exports, increased employment, and greater use of agricultural resources and infrastructure. Meeting the strict quality and the requirements of the United Kingdom market, may, however, be possible only for large scale producers or by the creation of export organisations to manage the marketing of small scale production. Some developing countries have been most successful in fruit and vegetable exports. For example, following rapid growth, exotics are now Kenya's fourth biggest export earner, and a further ten per cent growth is planned by the year 2000. Kenya earned KSh17 million from avocado exports and KSh50 million from mango exports in 1986.

The rates of growth in consumption of some exotic fruit are extremely high. Pawpaw imports have increased almost tenfold and mango imports have more than trebled in the last eight years. However the present consumption of exotics is still small. For example the best known exotic, pineapple, is only consumed at the rate of one quarter of a fruit per head.

Among other handicaps to developing countries are the market charges on the fruit. The mark-up at the supermarket is some 35 per cent and that of the independent green grocer is 45 to 60 per cent. On developing country produce there is a wholesaler commission which averages seven per cent as well as an importer commission which averages seven per cent. Both these charges and the mark-up has the effect of reducing the net price to the producer. This is before protectionist policies of the EEC are taken into consideration. The object of the study was to provide information which will assist developing countries exporters, actual or potential, in identifying profitable trading opportunities within the market, and also to give some guidance for the design of their production and marketing strategies such that foreign exchange earnings are maximised.[6]

Table 11. Estimated Retail Value of UK Fruit Market, 1986 £ million

Traditional commodities	£ million	Exotic commodities	£ million
Oranges	171.2	Avocados	21.7
Bananas	243.4	Mangoes	19.8
Apples	397.1	Pawpaws	0.8

Source: Fresh Fruit and Vegetable Information Bureau 1988

Table 11 puts into context the small but rising market for exotic fruit. Supplies of exotic fruit arrive in the UK from all over the world. Certain countries, like Israel, South Africa and perhaps Brazil have already taken a strong lead in the market.

The study was focused on six fruits, and in particular on avocados, pineapples, mangoes and pawpaws. The six fruits are broadly representation of the cross-section of the range of exotic fruit in the UK market. The characteristics of exotic fruits are as follows. Firstly, they should have an exotic,

---

[6]The UK Market for Fresh Exotic Fruit, Overseas Development and Natural Resource Institute, 1988

usually tropical or sub tropical origin, although avocados and kiwifruit, for example, are produced in more temperate regions including Europe. Secondly, they must have an exotic nature, for instance flavour or scent, unusual appearance in terms of shape, colour or texture, sometimes fragile and highly perishable. They should be unfamiliar to the majority of the indigenous population, although they may be part of the every day diet of ethnic minority groups, and perhaps not included in fruit retailers traditional stock. Exotic of course, is a dynamic quality and successful exotics may eventually become commonplace at least in terms of familiarity and availability. These characteristics give rise to the basic marketing problems associated with exotics. Their distant origins mean that careful packaging and transportation are essential. For many fruits, because of their perishable nature, air transport is the only means by which they can reach UK in an acceptable condition. High transport costs imply high price and exotic fruits are only bought by the relatively wealthy. The importance of building up the market through sales promotion and consumer and retailer education, is also stressed by the report on the UK market for exotic fruit. There is a low level of fruit consumption in the UK and overall demand for fruit in general is static. The demand for stone fruit, easy peel citrus and, in particular, exotics, has been increasing. The most significant feature of these fruits is their apparent rapid growth in demand. In the UK, between 1978 and 1988, pineapple imports increased by 151 per cent, mango imports by 367 per cent, avocado by 202 per cent and pawpaw imports by 1740 per cent. In the trade, most in demand are avocados, pineapples and kiwi fruit followed by pawpaws and mangoes. Expansion of demand for exotics seems certain to outstrip that for fruit in general and the market for the existing exotic fruit will increase, especially for pineapples, mangoes and avocados. However, the competitive position of many developing countries is weak. Exporters often lack information on the high marketing standards, in terms of consistency of quality and the regularity and reliability of shipments which must be met. Exporters often lack information on market structure and the best marketing channels, so developing countries may fail to maximise resources. Export organisations can play an important part in facilitating trade development and personal visits of importing companies to sources are valuable. Export organisations in the producing countries can provide a range of services to facilitate and co-ordinate exports. Technical advice, training and support to producers, a legislative and administrative frame-work to cover aspects of quality control and packing, and the representation of producers' export interests. Storage at airports and ports is important. Information on the export market requirements, prices and trends, and advice on export markets and the most profitable segments within them is also important. The assistance of the diplomatic service in the purchasing countries concerned is also of great value.

There has been a fundamental change in the food retailing structure in the UK during the past ten years. This has led to the emergence of a small number of national supermarkets who now have almost half of the produce market, a market share which they are trying to increase. Their system of marketing has had long term consequences for the produce industry and for all those involved, producer, distributor and consumer. As a consequence of income growth and improved standard of living in Britain, spending on fresh produce is increasing. Fresh produce in general and fruit in particular, have been beneficiaries of the changing market environment. Consumption of fruit has started to grow, not only in volume but also in value. In the distribution of fresh fruit from 1976 to 1987 the supermarket share increased from 22 to 49 per cent. The greengrocers share fell from 37 to 28 per cent and market stalls fell from 25 to 17 per cent. The increasing market share of the supermarket with sophisticated customers and superb selling, will mean that more of the exotic fruit and vegetables from the developing countries will be sold in Britain in future. The supply of fruit and vegetables is becoming increasingly channelled through large and sophisticated companies involved in all aspects of importing and distribution.[7]

The market for exotic tropical fruit is estimated as being worth £66 million at the wholesale level. There have been two underlying forces which have stimulated the growth of exotic fruits. One is the presence of three million ethnic people in Britain. This gives importers and retailers in Britain an opportunity to develop new products from new sources of supply. Another influence has come from the practice of a significant amount of the people in Britain taking cheap continental holidays. A wide section of the population have now experienced foreign travel and have enjoyed a range of unfamiliar

[7]The UK Market for Exotic Tropical Fruit, Henderson, D.W., Aspects of Applied Biology No. 20 1989

fruit and vegetables. These products have started to be eaten in the home and in restaurants on special occasions, and they slowly become part of the regular diet. Both these market forces have assisted in the growth of demand for exotic fruit and vegetables. The exotic fruit and vegetable market potential is great in Britain and other West European countries.

The EEC Protectionist Policies for Fruit and Vegetables as a Handicap to Developing Countries

The handicap for developing countries in the fruit and vegetable trade over the world, is the protectionist policies of the main world markets. That is the European Economic Community, Japan and the United States. Japan has the highest effective tariff rates in both categories. The European Community has the second highest rate for fruit, with the United States second for vegetables. Some developing countries send an overwhelming proportion of their fruit and vegetable exports to developed countries. The trade restrictions are of crucial importance to the expansion of developing country exports. As we have seen, tariffs on fruit and vegetables vary by product, season and country of origin. Protection is often higher in the home season of produce of the developed country concerned. At the same time a number of developing countries have beneficial agreements with the country or countries, to which developing countries send their exports. For example, the United States with the Caribbean countries and the EEC with the African, Caribbean and Pacific (ACP) countries. EEC trade policy for fruit and vegetables is given in Chapter 3.

Non-tariff barriers to trade

A large range of non-tariff barriers also applies to the trade of fruit and vegetables. They include quotas, voluntary export restraints, variable levies, minimum import price systems, countervailing duties, technical specifications, restrictions on labelling and packaging specifications. Of special importance are the plant health restrictions.

Data from UNCTAD indicates that in industrial countries, 39 per cent of fresh vegetables and 20 per cent of the fresh fruit is affected by non-tariff barriers.[8] The restrictive effect is considerably greater than that of the tariffs, as indicated by the price spread between domestic and import prices created by non-tariff barriers. For example, according to Ishon, the differentials between the import price and the domestic price, due to non-tariff barriers on fruit and vegetables, ranged between 30 and 80 per cent, whereas the corresponding price differentials due to nominal tariffs ranged from between 7 and 13 per cent. For fruit, price differentials due to non-tariff barriers ranged from 22 to 180 per cent and for nominal tariffs, from 4 to 14 per cent. All importers of fruit and vegetables maintain a system of orders that serve to protect and regulate their domestic production. Quotas and voluntary trade restrictions are also used in the fruit and vegetable trade. The liberalisation of trade barriers by developed countries would greatly increase the quantity of world exports.

Tariff and non-tariff barriers are a great handicap to developing countries. It must be contended too, that one of the major barriers to developing countries is their poor infrastructure and the chances of getting their business to be competitive, be it a producer association with a cooperative or a private individual. Producers, even in a developed economy, are not sure to be successful fruit or vegetable producers. In a developing country, it is more than likely that education is at a very poor level, which is a considerable handicap to producers in developing countries. For the developing country producer to succeed, he needs technical information about the fruit and vegetable enterprise and the capacity to implement it. To begin with he needs to understand the technical information. As stressed before, the role of government or a specialist market agency, with government backing, is most important in the training of the developing country producer.

---

[8] Horticultural Exports of Developing Countries, Past Performances, Future Prospects and Policy Issues. Nural Ishon Islam- International Food Policy research Institute, 1990

The Prospects for Developing Countries in the European Market for Fruit and Vegetables

The prospects for developing countries in world trade in fruit and vegetables are promising, especially when consideration is taken of the rising demand for exotics in the developed economies. These prospects depend on the level of disposable incomes in the developed world which in turn depends on their economic success. This means that the benefit of trade in fruit and vegetables in the developing world is directly linked with economic advance in the developed country economies.

Two principal features affect the opportunity for developing countries in world trade in fruit and vegetables. These are disposable incomes in advanced economies which represent the market for developing countries' produce, and the protection of fruit and vegetable given by developed economies. Rising affluence in the developed countries will assure a bigger market for developing economies' production of exotic fruit and vegetables and if the developed countries are persuaded to reduce protection this will help the developing countries' exports too. If protection is maintained at its present level the prospects for developing countries exports are not so good. For protection to be reduced or eliminated, the prospects for developing countries would be much better. However, there is expected to be a regular demand in the Western economies for developing countries' produce, with more emphasis now on the more exotic fruits. At the same time, the large world trade in bananas illustrates the great importance of traditional fruits and vegetables in the exports of developing countries. Developing countries have a definite stake in the market for exotic produce now and they are likely to improve their position in the future.

Table 12. Bananas in World Trade, 1988, thousand tonnes

	Exports		Imports
World Total	**7885**	**World Total**	**7610**
Developing countries	7445	Developing countries	505
UPEB countries	3845	Near East	168
Other Latin America	1761	Developed countries	7105
Ecuador	1535	Western Europe	3121
Caribbean	617	EEC	2660
Africa	187	E. Europe and USSR	201
Asia and Oceania	1034	E. Europe	135
Philippines	867	USSR	66
Developed countries	440	Canada	230
Spain (Canary Islands)	400	United States	2750
		Japan	760

Source: FAO Commodity Review and Outlook, 1990

The EEC Banana Sector

The banana is the most popular fruit, in the EEC, after apples and citrus fruit, while in the United States they are the most popular fruit commodity. The main producing countries are located in Central and South America, as shown in Table 12. The Caribbean and West Africa are also important areas for banana production. Within the European Community, the banana is only grown in Crete, and it is also grown in the overseas territories of a number of Member States - The Canary Islands (Spain), Martinique and Guadalupe (France), Madeira (Portugal). The Economic Consultants in

Dublin have produced a comprehensive report on bananas and the EEC market, which shows how diverse the banana market is in the EEC.[9]

Bananas are among the largest export crops of agricultural commodities. The total bananas exports in 1988 has been put at 7.9 million tonnes. Production has risen from 1985 to 1988 by 13 per cent, and 95 per cent of bananas comes from developing countries. The world exports and imports of bananas for 1988 are given in Table 12.

Table 13. Bananas in EEC Trade, 1988

	Imports '000 tonnes	Imports per head
Member States	2687.4	
Belgium-Luxembourg	86.0	8.6
Denmark	34.0	7.6
France	454.8	8.7
West Germany	757.0	12.3
Greece	3.0	0.3
Ireland	22.0	5.9
Italy	380.5	6.6
Netherlands	126.5	8.4
Portugal	78.6	7.5
Spain	360.0	7.5
United Kingdom	388.0	6.9

Source: FAO Commodity Review and Outlook, 1990

The European Community is the world's biggest market, larger than the market in the United States and Japan. Total consumption in the EEC has risen by a quarter in the last five years. The largest EEC market is West Germany followed by France, the UK and Spain. Imports and consumption per head for the EEC is given in Table 13. Belgium-Luxembourg, Denmark, France and the Netherlands all have a consumption per head of seven or eight kg per week. West German consumption of bananas is much higher than that of other member states, by a very large margin. Consumption in West Germany is over 12 kg per head. Some quarter of EEC supplies of bananas comes from the French overseas department together with Mediterranean Islands and Crete. A further 20 per cent comes from Africa and the Caribbean (ACP) countries. The balance of supplies of over 50 per cent come from the dollar zone. The banana markets of member states divide into a number of groups. The UK, France and Italy protect their markets in favour of former colonies now in the ACP. Greece, Spain and Portugal protect their markets in favour of their "domestic producers". The remaining EEC countries operate liberal open market systems, and their bananas come mainly from the cheaper dollar zone. West Germany, with practically twice the consumption of other EEC countries, has not imposed a duty on banana imports for a number of years.

So the present EEC banana market involves a variety of national arrangements. The six member country nations, mentioned above, regulate their market in the interests of either preferred overseas suppliers or their own producers. This means that intra EEC trade is disrupted in order to

[9] Trade Policy and the EEC Banana Market, Bowen Brid, Fitzpatrick, James and Matthews, Alan, Economic Consultants, Dublin 1990

prevent cheap dollar imports leaking on to the protected national market through the liberal member states. Article 115 of the Rome Treaty is used extensively in this regard. The net effect is that low cost dollar bananas are excluded wholly or partially from the national markets of half the EEC member states, in favour of more high cost ACP or domestic sources. With the exception of West Germany, all larger member state markets are regulated in this way, constituting over 60 per cent of total EEC banana trade.

The EEC has no common external trade policy for bananas. A network of traditional practices in the banana trade in individual states continues to operate. This includes a series of bilateral preferential and protectionist trade regimes. The Community's aim for a single market by the end of 1992 for goods and services, will have far-reaching implications for the present diversity of the banana trade. The existing situation is not likely to survive the thrust towards increased internal market liberalisation and the external market harmonisation which 1992 marks. The Uruguay round of GATT negotiations, with its emphasis on increased agricultural trade liberalisation, adds further impetus to these trends.

A variety of different proposals for post 1992 banana trade has been made. In general, these involve various combinations of price based import restrictions, (tariffs and levies) and quotas. The nature of their application ranges from a liberalised EEC internal and external market on one hand, to a harmonised protected market on the other. "Liberal" and "protectionist" here, refer, in particular, to the dollar banana perspective, since the purpose of a Community protected market is to protect the higher cost Community and ACP produce.

Implications for Fruit and Vegetables of the Single EEC Market 1992

The implications for the single market for bananas have been discussed in Chapter 8. While there are, now, few barriers to the free flow of fruit and vegetables within the EEC, the Single Market is expected to remove all remaining barriers. At the present time there are a number of irritating administrative obstacles to the free flow of produce, such as export licences, and phytosanitary and plant health certificates. It is proposed that these will be eliminated or at least streamlined by 1992. A contentious area is the validity of the phytosanitary and plant health certificates which accompany imported produce. Inspection schemes, grading standards, labelling and price reporting are now almost standard throughout all member states of the EEC. Further small modifications are expected before 1992 but the effect on the fruit and vegetable sector will be minimal.

The operation of the CAP for fruit and vegetables and changes in the regime for this sector will continue to override the effects of the Single Market programme so far as the production of fruit and vegetables is concerned. The regime effectively restricts competition from imports and protects a single market. The EEC market for fruit and vegetables is already highly competitive, and the completion of the single market will ensure absolutely free competition, as the remaining restrictions on trade are removed.

The harmonisation of plant health regulations will greatly facilitate the movement of produce between states and will reduce the costs of meeting plant health standards. Producers will have to satisfy only one set of standards in order to trade freely across the Community instead of the numerous and different national standards operating at present. The establishment of the Single Market will increase the availability and the choice of financial services for producers. Advantage will be taken of the wide range of credit and loan facilities specifically designed for the agricultural sectors in other member states.

The removal of restrictions regarding transport and the elimination of border checks should be of great significance to the trade sector. Delays at border crossings are responsible for losses where consignments of perishable fruit and vegetables are transported to markets in other member states. The time-saving from the elimination of the restrictions will make better use of transport. The cumulative effect of the Single Market is likely to give increased competition in both the domestic and overseas market. Domestic market competition is likely to increase because of an increasing awareness of the subject of 1992. The completion of the Single Market will be of great significance to distributors as it will remove restrictions on transport within the EEC. There are no trade barriers between the member states in the former Community of 10. Spain and Portugal have special transitional arrangements to be phased out by 1996. A minimum import price system will continue to apply to Spanish and Portuguese exports to the ten for certain fruit and vegetables at sensitive times of the year until 1996.

The single market will not change the quality standards which apply to some 30 different fruits and vegetables, notably on the labelling requirements of the standards, which will continue to require the origin of produce to be shown. The quality standards apply to all steps of distribution and apply to both home-grown and imported produce. The marketing Inspectorate in each member state will continue to enforce standards. It is not yet clear whether export certificates and border checks will be maintained at current levels and format.

INDEX